I0065432

Homo Falsus – Lying Man

Homo Falsus – Lying Man:

Physics, Consciousness, Language, Information, and I

Boris Rusakov

Copyright © 2023 by Boris Rusakov

All rights reserved.

No part of this book may be reproduced or used in any manner without written permission of the copyright owner except for the use of quotations in a book review. For more information, write to: rusakov@xpertnetinc.com.

Publisher: Xpertnet Inc
www.xpertnetinc.com

Editor: James Kingsland
Book design by Kristina Eraie

ISBN 979-8-9882910-0-8 (hardcover)
ISBN 979-8-9882910-1-5 (paperback)
ISBN 979-8-9882910-2-2 (ebook)

Printed in the United States of America

To my daughters Adina and Miriam

To my wife Alona

*To memory of my mom Sara, my aunt Gesya (Galya),
and my dad Efim (aka Haim or Fima)*

Contents

1. Introduction. Who is Homo falsus?

Homo falsus is a deceitful, mistaken, or lying man. For some time now, this is what I have called the species to which I belong, and I hope you will too after reading this text. One could call us *Homo loqui* ("talking man"), but speech is merely a means of transmission, while falsehoods (or to be more lenient, abstractions) are the essence of what we convey by speech. That is, lying is the cause and speaking is merely the effect. Language emerged as a means of denoting and transmitting falsehoods. Below I will show that the lies that we tell whenever we open our mouths and turn on our speech organs are completely innocent, though not necessarily harmless. Strictly speaking, they do not have the purpose of misleading anyone at all, though this is their inevitable effect. In any case, the first person to be deceived by these lies is the speaker himself. Therefore, we might as well be called *Homo decepta*, a "deceived man". Once again, the "lies" that we are discussing here are unconscious lies. They are not like the deliberate lies used by professional liars – politicians, journalists, and others –who earn a living by lying. We are talking about the unconscious lies of ordinary, normal, mostly honorable, and decent people, who are the majority. And that's what I'm here to try to explain, among other things. We think we are telling the truth, we think we know it, but in fact we have no idea what we are saying. If, after reading this text, you still have an unpleasant feeling that you were offended by being called a liar, you can call yourself an abstract thinker, a creative inventor, a dreamer, or something similar, although they are all the same.

Why don't I like the name *Homo sapiens* (a "wise, understanding man")? Because it's another deception. It is fundamentally wrong, and means exactly the opposite of who we are. We think we understand something. In fact, we are wrong about almost everything. It's true that the nonsense we constantly spew is what fundamentally distinguishes us from our animal ancestors. But if you really want to flatter yourself, you may call this faculty "abstract thinking", or "inventing something that does not exist". In this book, I want to point out that our arrogance, our love for ourselves, and the desire to attribute non-existent qualities to ourselves, are very harmful to us. These self-deceptions prevent us from understanding ourselves, our consciousness, our origin, and many other important things about ourselves and the world around us. I deeply believe that now is the time for us to acknowledge this if we are to continue exploring the world around us and preserve ourselves as a species. Our entire history, and even more so recent events, clearly show that our survival not only cannot be guaranteed, but often hangs in the balance, so that only a lucky chance separates us from total extinction. In this text I will reiterate often my plea to refrain from self-aggrandizement and examine ourselves honestly.

For starters, I ask you to at least come to terms with naming yourself *Homo falsus*, a lying man. Notice the leniency. I didn't suggest naming us a lying or talking ape, although that wouldn't change the essence of who we really are. If you doubt the ape part just look in the mirror. This is also evidenced by our comparative analysis of DNA. According to various sources we share around 99% of our genes with our closest animal relatives, bonobos and chimps [1].

As is clear from the subtitle, this text will be about consciousness. Why did I, a former theoretical physicist, an author of a couple of pretty (without false modesty) formulas in two-dimensional quantum chromodynamics that are very remotely related to the real world, being neither a biologist, nor a psychologist, nor an anthropologist, decide to delve into the subject of consciousness, and even write about it? Of course, like anyone who wants to know something about themselves and the world in which they live, I have always been interested in this subject. And certainly while doing science, I could not help but wonder how my own consciousness works, at least with the aim of making it work better. I left science and moved into business a long time ago, and since then have had no time for it. A few years ago, having gained a certain amount of free time, I decided to finally find out, i.e. read about, what consciousness is. I thought that after so many years science would have already figured it out. It is not for nothing that everyone is now busy with artificial intelligence. Therefore, I assumed that natural intelligence should have been long understood.

But alas, it turns out that despite the huge amount of recently accumulated empirical material [2], a theoretical understanding of consciousness is almost completely absent. Instead of consensus, there are fierce debates and battles between the various schools of thought. Their authors and followers, all of whom are brilliant minds, are unable to describe what consciousness is. The neuroscientist Tatyana Chernigovskaya, in her exciting lectures, poses a lot of intriguing questions, but does not offer any answers [3]. The brilliant philosopher Daniel Dennett echoes these questions in many ways and claims that consciousness is an illusion, but this is an answer that will hardly satisfy anyone without an explanation for exactly how this illusion works [4]. David Chalmers, a no less brilliant thinker, says that this is an almost unsolvable problem [5] (the "hard problem" of consciousness). And so on. I hope numerous ancient and modern thinkers forgive me for not citing them. The reason is that I never found a single text, including those quoted above, giving an adequate (from my point of view) explanation of what consciousness is, where it came from, why it exists, how human consciousness differs from an animal's one, and how this difference emerged. If I had found it, you would not have this text in front of you.

So I decided to think for myself. To my surprise, after some time of comparing and re-thinking the facts known to all[1], I began to understand something. And after some time I began to write

---

[1] Known to me, a layman. Therefore, I consider them common knowledge. It is clear that professionals know much more.

down what I understood, in order not to forget it. Today, I'm not afraid to say, I can probably answer all the questions asked by Chernigovskaya, explain to Dennett what the "illusion" is, and prove to Chalmers that there is no hard problem. Everything is amazingly simple! Simple and beautiful. As it should have been. And that is what I want to tell you about.

The key to understanding consciousness for me was the definition of information and the inescapable conclusion that it is made up of concepts. The very concepts we are all familiar with. The key was to realize that concepts are **everything** that our consciousness (and information) comprises. We often use this term when referring to a scientific concept. But it requires some effort to realize that such simple notions as "mom" or "dad", "sit" or "stand", "tree" or "bush", "I" or "you", are in fact concepts, too, in other words abstractions. It is not difficult to understand that in order to "receive" any information you need to have a concept of such information. I put "receive" in quotation marks because this is another deception. We do not "receive" information, we produce it [6, 7]. Both concepts and information are covered in chapters 4, 5, and 6.

The second ingredient that helped me understand what consciousness is was thinking about why it is needed at all, where it came from in the sense of how it evolved, and thinking about nothing less than the "meaning of life" – not in a poetic sense, but from the point of view of nature. Also, in order to understand consciousness, it is critical to realize that consciousness and vision are (almost) the same thing. After we have defined vision in more detail, we can remove the word "almost". This is covered in chapter 10.

Since the summer of 2022, I have been trying to partially publish in various "scientific" journals the most relevant things that I have learned [6-8]. But not so fast! My attempts were furiously rejected not only by religious Shannonists, whose positions and salaries seem to be tightly linked to the so-called Shannon Information (although Claude Shannon himself did not give any definition[2] [9]), but also by journals claiming authority in the field of consciousness, whose referees just didn't grasp anything from what I wrote. One of them couldn't even hide that he didn't understand my definition of concepts. Some editors politely rejected my articles on the grounds that they did not correspond to the profiles of their journals (which was impossible to guess from their titles and declared scopes). One editor informed me that he could not accept my paper because it did not present newly discovered experimental data (as if all the old data were perfectly described by their non-existent theories). Apparently, this journal exists to publish reports on how taxpayer money was spent, while inexpensive rethinking of facts is not considered real science. It seems that the dominance of pseudo-scientists in these areas is much deeper even than in my native theoretical physics, which I left not least for this reason, and that in order to publish there one has to belong to one of the established schools of thought (or sects, or mafias), explain where you stand in relation to Aristotle, Descartes, and thousands of subsequent deservedly famous thinkers, and not just walk in from the street and tell them how simple everything is.

In the end, I returned to where I started – to my original notes – and decided to turn them into the text that is now in front of you. I thought it might even be for the better, since this format allows me to present the whole picture in an informal yet more coherent way. What came of it, you must judge for yourself.

---

[2] The word "information" in his article could easily be replaced with "data" or "message" without any harm to the content or significance of his work. Its use was disputed in [10], and a further history of the discussion can be found in [11].

Just a few words about whom this text is for. As I said, I originally wrote it for myself. So it was pretty short. I realized a long time ago that writing long texts is not my thing. My old articles on physics were very short, 10-15 pages maximum. But that was mathematics and physics, and their main language was math formulas. The subjects of this writing require extensive explanations in plain language. However, our "plain language" is grossly inadequate to this task, for the reasons I explain here – because of the deceitful nature of language itself, and especially when it comes to everything related to consciousness. Realizing that now I would have to present this text to the public, I began expanding it, because what took me months of deliberation can hardly be conveyed in a nutshell, without examples. So I give as many examples as I can to illustrate my sometimes seemingly unfounded statements. I will nevertheless be quite brief, and still rely on thinking readers to ponder what I am saying, and to come up with examples to support (or refute) these assertions, if they do not think that I have sufficiently supported them. Primarily, this text is intended for researchers, physicists, mathematicians, philosophers, and all those who are not alien to scientific thinking. However, I believe that any thinking person interested in the subject of consciousness and its origins can read this text. Occasionally I refer to physics and use its terms without further explanations. Consider this as a bonus for physicists. If you are not a physicist, you can simply ignore such references and keep reading, as they are not crucial for general understanding.

A few words about the structure of my presentation. As you will see, consciousness, information, language, and other subjects I cover are based on concepts, and concepts in turn are based on sensations and feelings. At first, it may seem that all this is a rather twisted puzzle. However, this is only the first impression. To me, the beginning of unraveling this puzzle was the definition of information, which led me to concepts. In turn, the realization of the role of concepts as elementary constituents of human consciousness provided an understanding not only of the functionality of consciousness, but also of its origin.

As I worked on this text I changed the order of presentation several times. In fact, for the discussion of consciousness, information is a side issue. It could be left aside, discussed separately, or even excluded altogether. However, in the end, I decided that if you follow the same process as I did– starting with information – it will be methodically beneficial. The only deviation from my personal history of events was to add two small chapters at the beginning. In the first of these sections (below) we will discuss sensations, which are the physical, material basis for everything we will be discussing, while in the second (chapter 3) we will examine language as an instrument of (self-) deception. Next, we'll jump straight into a discussion of where information is born (chapter 4), but before we can define it (chapter 6), we'll have to take a break and talk about concepts (chapter 5). After that, the presentation is quite continuous.

2. Motion, inertia, and sensations

For a reason that I will touch on in the last chapter, we exist in a world of events. An event is a change in something, at least one parameter. We can say that we live in a world of motion, if we interpret motion as any change, and not just as a change in location. However, the word "live" already means motion. Without motion, there is no life. Motion is perhaps the only fundamental property of the entire world we exist in. Any motion results in compensatory motion, counteraction, or inertia, which arises from the fact that our world of motion is part of a motionless world. Assuming that this is so, the motionless, stationary world is not affected by the motion of our world, because this motion is completely compensated by its counter-motion. This is why we have Newton's Third Law, all kinds of conservation laws, and all the so-called "forces of nature" and "laws of nature". These will be discussed in the last chapter 13.

The existence of the entire world around us would be impossible without motion. Life itself, the possibility to observe this world and to gain knowledge of it would also be impossible without motion. In order to learn about the existence of something, a signal needs to be sent and/or received. To measure, you need to make an impact – to interact with the subject. We can only measure motion, changes in parameters, differences, but not absolute values. Without motion, consciousness would not exist either. All our sensations are based on the reactions of our nervous system, or brain. The reactions are compensatory, inertial responses to external perturbations, to the motion of signals, neurons, molecules and atoms.

Our feelings and sensations are interpretations of the nervous system's reactions. They are therefore the basis of consciousness, information, concepts, and everything that is discussed here.

The vast majority of the reactions of our body to what is happening in it go unnoticed by us. Actually, it would be more correct to say "we don't get informed about them". More exactly, these reactions remain uninterpreted. As I just explained, from a physics point of view, any motion produces a counter-motion, a recoil that cannot but be "felt" by neighbors. Fortunately for "us", "we" do not flinch at every exchange of signals between our neurons. We do not feel chemical reactions or macro-processes, such as the blood flowing in our veins and capillaries, or the heart beating (at least if everything is fine). We hardly even notice our own breath. Imagine if that were not the case. Our lives would be a complete nightmare.

We are only "told" if something is wrong. Then we begin to feel pain of varying degrees or discomfort. So someone is still watching this. Moreover, he is very carefully and constantly watching. Not only does he observe, he also heals us if something is wrong (which you may only notice later by, for example, your elevated body temperature). Yet, you do not include this "healer" in your "I" because he is too lowly to qualify. You attribute only top functions to your "I", such as thinking about the eternal or scratching your heel. It's good that he does not bother us over trifles, does not interrupt our sleeping or "thinking"[3].

What I want to say is that the very fact that we do not notice something, or do not feel it consciously, does not mean that nothing happens or that someone else in our body does not sense it. There is someone who senses and observes carefully. He just does not inform us.

---

[3] I apologize for the episodic multiple quotes. To be completely honest, every word here should be in quotes. I have so far quoted only the most outrageous lies. We have no idea how and who thinks (see below).

We have divided all our senses, or sensations, into five, simply because we have counted five external sensors that are noticeably different from each other – eyes, ears, tongue, nose, and skin. In fact, external signals make up a very small fraction of what our brain (or consciousness) processes. Most of them are produced in the brain itself, and the number of sensors is infinite. Besides billions of neurons, our brains also contain neural networks. These are clusters of neurons that not only have various functions, but also change their functions constantly and acquire new ones.

3. The problem of consciousness. Language as a tool of deception

To understand what consciousness is and how it works on a functional level we will have to redefine many familiar concepts such as information, I, thought, and many others, because our understanding of their meaning has so far been the result of deceiving ourselves.

Judging by the definitions I found, this deception is global and comprehensive. It is firmly rooted in our language, and we deceive ourselves every day, every second, in everything that has to do with consciousness. To some extent, I am ready to argue that deception is one of the main functions of language. It is deceiving not only others, but also, and primarily, ourselves.

Another reason why consciousness is so difficult to study is its privacy. Not only can we not penetrate each other's thoughts, we cannot even determine such simple things as, for example, whether we see the same color in the same way. I would love to see any color "with your eyes" at least for a minute. Maybe what I perceive as green, you view as blue, and vice-versa? It is easy for us to determine that a color-blind person is such, since he cannot distinguish one color from another. But it is impossible to discover exactly HOW another, non-color-blind person "sees" (or better to say "feels") a specific color[4]. I will periodically return to this issue, and I hope that I can demonstrate the impossibility of "reading" other people's thoughts, since they are nothing but sensations.

Our language is not only inadequate for describing everything that is related to consciousness, but is the main tool of deception. When we use words such as "I", "thought", "consciousness", "think", we believe that we know their meaning. You will see very soon that this is not so, and that it would be correct and fair to put every word in this text in quotes. To make reading simple, I will, of course, take pity on the reader and refrain from doing that.

Let me start with a question. How do you think? No, I'm not seeking your opinion on something. I ask how exactly you do perform the process of "thinking". If, like me, you don't know how you do it, then let me ask you the next question. Why do you claim that you do something if you cannot even remotely describe how it is done?

It was a long time ago when my youngest daughter started learning math at school. She regularly came to me ostensibly for help. However, I quickly realized that in fact she was looking for a ready answer, because she did not want to think for herself. One day, when she approached me with a problem that in my opinion she could have solved on her own, or at least start solving, I said: "Think for yourself." She replied: "How do you think?" which flummoxed me. Seriously, how could I ask her to do something that I had no clue how to describe doing? We can only provide motivation, and hope that a person will somehow learn the trick.

You can easily and in detail describe any of your actions that are not related to the work of consciousness. For example, how do you dig a hole? You pick up a shovel, stick it into the ground, tilt it and use it as a lever to lift the soil, then throw it. But you have no idea exactly HOW you think. Moreover, not only the average person has no idea, but also a neurosurgeon who has seen the brain, and a cartographer of the brain who knows where, what, and when it gets electrically excited, and a brain physiologist who knows how neurons communicate with each other, and a specialist in neural networks who knows how they arise and organize themselves. Even an artificial intelligence specialist, who knows how to program a machine to "think", has no idea how one thinks. Wouldn't it be easier to assume that the reason is that it is not "you" who thinks, but someone else? Meanwhile, "you" are just eavesdropping and trying to verbalize what you overheard.

Have you ever had a very interesting and profound thought, but a moment later, while you were about to verbalize it, it disappeared somewhere, and now you not only cannot remember the thought itself, you cannot even say what area it belonged to? So whose thought was it? Was it really yours? And then, who is "you"?

And how is it possible that you suddenly start thinking about what you don't want to think about at all? You were going to think about something important, to ponder some problem which you really wanted to solve. And then, instead, some completely extraneous thoughts "enter your head", some old memories, some old movie, or some long-forgotten annoying song that you don't even like to hear. And it takes a lot of effort for you to get rid of the unwanted thought and "switch" back to thinking about what you were going to "think". Often it doesn't work at all. Don't you think that someone else is thinking those extraneous thoughts? Someone who doesn't give a damn about what you were planning to think about. So who is this other one, and which one of you is "You", or "I"?

As you may already have begun to suspect: 1) "I" is merely an outside observer of what is happening in the brain, and not in any way its "owner" or even "controller", and 2) our language is totally inadequate for describing what is happening (but what else could we use?).

We were so successful in deceiving ourselves about consciousness that we now need to make a serious effort in order to expose the deception if we are to shed light on this issue.

I want to emphasize that despite using such negative terms as "liar" and "deceit", I fully appreciate the achievements of human intellect, consciousness, and language. They are worthy of admiration. Thanks to them we have our civilization, cities, books, roads, ships, atomic energy, cars, and airplanes. By the way, this generalizes a point made by Harari in his book [12] that the achievements of our civilization would be impossible if we hadn't invented states, nations, money, banks, corporations, and so on – all of which are fictions that don't exist in nature but only in our minds. And yet it's amazing how well consciousness works! It is amazing that you can use consciousness, and use it quite successfully, without caring at all about what it is and what happens in it. Even the fact that consciousness deceives itself in no way diminishes its achievements. So we will assume that so far it has been a deception for the good, though not all achievements of human consciousness have been peaceful. Unfortunately, having made a discovery, humans first of all try to apply it to the creation of new types of weapons and means of destruction or the subjugation of their own kind[5]. But it is also obvious that the time has come to understand how consciousness actually works, and perhaps, armed with this understanding, move further in exploring the world around us.

4. Where does information come from?

In our minds, we believe that our consciousness deals with the processing of information coming from outside, which is its primary function, perhaps the only one. Already on the next page you will see that this is not the case. Information does not "come from the outside", but is generated by human consciousness. Generally speaking, information is a separate topic, and I could easily leave it out. Nevertheless, it is very useful for understanding the difference between human and animal consciousness, and therefore consciousness in general. For me it was the turning point in untangling the puzzle of consciousness. Therefore, for methodological reasons, I decided to start my presentation with information.

Oddly enough, despite the rapid development of information technology, artificial intelligence, as well as the sciences related to the study of consciousness, the brain, and neural networks, it seems that no one can consistently define information. And this is despite the fact that information is the main subject of all of these sciences, and that we constantly use this word and intuitively somehow understand its meaning. However, the meaning varies from person to person and is dependent on the context. There is no consensus as to its origins or whereabouts, or about its units of measurement, or even whether it is a material entity. The IT specialist would tell you that information is a sequence of zeros and ones, and is measured in bits, a physicist would tell you it is measure of inverse entropy, a marketer or intelligence officer would say it is a dataset or knowledge about something.

Let's try to figure it out. To begin with, we would like to understand **where** information originates.

Among hundreds of definitions of information one can find

via Google, the following stands out because of its simplicity: information is what **informs**. It is nearly impossible to argue with this, no matter how stupid and tautological it sounds. The only question that remains is: whom does it inform? This definition implies some kind of recipient.

We used to think that information was contained in a message we received, in a newspaper, in a file, in a photograph– and generally in the outside world, regardless of the recipient. Every day, every second, we ourselves reinforce this delusion with the help of language: "I received information", "I was given information", "The message contains information", "The book contains a lot of useful information", and so on and so forth.

Now imagine this whole outside world without recipients of information. What is the meaning of all of these messages, files, photos, if no one can ever see, decode, or understand them? Do they differ from random artifacts of an inanimate nature? The answer is no, and there is no information in them. There is also no information in an encoded message that is never going to be decoded. If someone convinces us that those breathtaking shapes of Arizona Monuments are actually a coded birth certificate of John Smith from Alpha Centauri, we will rush to decode them. But if there are no recipients, then even a real birth certificate of a real John Smith from Kentucky carries no more information than the curves of the mountains. If no one is there who knows what it is and how to decode it, it's just an artifact of inanimate nature, no better than the peculiar shapes of some mountains.

Therefore, for information to exist there must be a recipient who has a concept of such information. The recipient has to know (or at least assume) what it is, what it is about, and how to decode it. In addition, the recipient must have concepts of the various components of the information. In the above example, in order to obtain information from a birth certificate, the recipient must possess the concepts of birth, date, name, gender, encoding method, and hundreds of others. If the recipient doesn't possess these concepts, he obtains no information, or obtains only the part of the information that corresponds to concepts that he does have. In addition, these concepts are a function of culture, language, and time. For example, my Soviet-era birth certificate contains the ethnic origins of each of my parents – a trick invented by the Russian-Soviet Nazis in order to single out Jews, just as their German predecessors did. On the other hand, I was surprised to see that in addition to relevant information, such as date and place of birth, an American birth certificate contains the weight and height of the newborn.

You know very well that two people talking to each other in a language that you do not understand are exchanging information, whereas for you their messages do not carry any meaning as you don't have a decoding device. Moreover, sometimes you don't understand your own language. Try attending a university seminar on theoretical physics or mathematics if you are not a physicist or mathematician. You will meet people who speak your language, and exchange apparently valuable information, but you get nothing. I experienced this shock when I attended my first such seminar as a third-year student. You seem to understand all the words, but cannot grasp the meaning of what is being said. So it's not just a question of language. It is a lack of relevant concepts.

If you spent most of your life in a city, or your home is located among forests, I guarantee you will be lost if you find yourself in the polar snowy desert. However, a local guide, having a quick glance at what looks to you like an impenetrable white plain, would instantly determine your location and a way out. For him, the slightest curves of snow hills carry a lot of information, while for you they mean nothing and you do not even see them. Your mind is not used to it, while his mind deals with it all the time.

What is information for one may not be information

for another. There are countless examples in which the same "information received", or rather the same signal, or message, produces completely different information in different people, and even in the same person at different points in time.

Therefore, if we attempt to define information as something contained in a signal, we have to admit that the signal contains an infinite number of "informations" at the same time. Thus, it is clear that defining information as something that exists independently of a particular recipient is just as meaningless as claiming that any cubic centimeter of space contains all the information about the entire Universe and every piece of it. You just need to be able to extract it. In a sense, of course, this is true. But in order to extract this information, the very same recipient is needed.

Thus, information cannot be defined as something objectively existing.

I reject possible accusations of being a solipsist. I do not deny the existence of an objective reality outside our consciousness. But information about this reality, unlike reality itself, requires the presence of a "recipient of information", one for whom this "information" makes sense. That's why I put quotation marks here, because there is no "information" outside the "recipient".

Information is produced by the consciousness of a "recipient". I use quotes because the "recipient" does not have to receive anything and is in fact an "information maker". What about the role of the sender and of the external signal? An external signal, interacting with the consciousness of the recipient, can cause a response that leads to the production of information. However, it may not cause any reaction, as we saw in the above examples, or, even if it causes a reaction, it may not result in any information at all.

In other words, even a very "informative" external signal may not produce information. In addition, information can appear even without an external signal. For example, if you were waiting for some event, but it did not happen, or a signal you were expecting did not arrive, is it information? Certainly. I assure you that if the sun does not rise tomorrow, it will create an extraordinary furore, an informational bomb. Of course one can say that the absence of a signal is a signal of absence. As a result, it would be even more evident that information is born in consciousness, since the interpretation of the absence of a signal as a signal of absence can only be made by consciousness. If consciousness has learned in the course of time that a certain signal has a time pattern, i.e. it comes with a certain periodicity, then the non-arrival of such a signal will be perceived as new information and subjected to immediate analysis. Thus, the expectation of a signal is part of the template that consciousness associates with this information. A break in the pattern is a template change that must be interpreted.

Not only is information a product of human consciousness, but the very ability to produce it is a sign of human consciousness. The fact is that in order to produce information from an external signal (and even more so from its absence), the recipient must be 1) trainable (capable of learning), 2) able to conceptualize and interpret. The first is a property of animal consciousness, and both together are properties of human consciousness. Let's consider learning ability in more detail.

The same external signal or message we receive during our lives produces more and more information in our minds over time, while in early childhood it did not produce anything. What happened? We have learned. The first time we received the signal, it didn't give us any information. But we remembered, and the next time, we discovered that the same signal produced information, and the more we received it, the more information there was. If we were not able to learn, each new reception of a signal would have the same effect as the first time, and this would not change throughout life.

The ability to learn presupposes the presence of memory, the ability to recognize, as well as the ability to improve one's reaction to subsequent receptions of the same signal. All animals possess this ability to varying degrees.

However, learning alone is not enough to produce information. It is necessary to be able to conceptualize and interpret the signals received.

5. Concepts

A concept is an abstract entity created by our consciousness and existing only in consciousness. It is a common property of different observables, their common classifier, a notion that corresponds to a specific set of observables. The most significant property of a concept is its independence from a specific object from which we single it out, its portability to other objects. For example, "color" is a concept that unites "red", "yellow", "green", "blue", and so on (although they themselves are also concepts). "Red", "yellow", "green", "blue" are *values* of the *concept* "color". Food is a concept, while meat, milk, and bread are its particular values. A tree is a concept whose values include oak, birch, spruce, etc. What is a concept in one case can be a value in another concept.

Conceptualization is the first sign of abstract thinking. Actually, abstract thinking and conceptualization are the same thing. Concepts do not exist in nature. They are only in the human mind, and are acquired in the process of learning. In addition, even in human consciousness, they are not present at birth and begin to appear much later, after some period of training.

Observing animals, it might appear that they also have concepts. Judging from the reaction of a dog to other dogs, to its owner, or to cats, one could assume that it has the corresponding concepts: "dog", "owner", "cat", "friend of the owner", "stranger", etc. However, this is only an appearance. What we see as a dog's reaction is just the result of its trainability. After an animal learns something in a particular setting, it can only apply the "knowledge" in the very same setting, but is unable to do it in a completely new situation. A concept, on the other hand, by definition exists outside of any object. It is the result of a process of interpretation and abstraction of which dogs are incapable. So what we see as a dog's reaction is just the result of its trainability.

21

Animals cannot extract a concept from an object and apply it to a different object, in a different situation. For example, you can teach your dog to give you a paw, but it will not give a paw to another dog. It is we who interpret the reactions of the dog as a sign of the dog having concepts, which is another illusion of our human consciousness.

One of the most critical properties of a concept is that it cannot exist without notation, which in our case is language. Any original concept must be denoted by a word or phrase. Without notation, the concept does not exist. A concept's name is an integral part of it and allows it to be transferred from one object to another. Certainly, language is not the only means of notation. It could be (and apparently was before the emergence of language) an image, a drawing, a meme, a sign, a sound, or similar. Nevertheless, language is the most versatile notation of all, and we are fortunate to have it.

In [12] Harari correctly pointed out that such concepts as money, religion, nation, state, the borders of countries, banks, and corporations do not correspond to anything at all in nature. They are human inventions. My assertion is that not only these but **every** word, every phrase we use refers to a concept created by a human. Even "mom" and "dad", even "apple" and "tree", and even the very word "even", are concepts, even though an apple, a tree, or parents are not fictions.

Let's take a closer look at basic, rather "real" concepts, and start with the concept of "mom". After birth, while you still did not know anything, you got used to the fact that there was a woman calling herself "mom" who for some reason that you didn't know took care of you, fed you, nursed you, etc. Then you learned that other people around you also had mothers, and even your mother herself had a mother (your grandmother). Thus, mom became a concept separated from a specific woman you knew. Then you found out that it was this woman who gave you birth, gave you life (together with dad). By this time you already had the concepts of life and birth. Thus, the concept of mother was constantly expanding and deepening. Then you realized that the woman you called mom was not necessarily your biological mother who gave you birth. The only way for you to know that she gave you birth was to believe it, to believe other people, documents, or DNA analysis. It also turned out that there was only one mother, two grandmothers, four great-grandmothers, and so on. And these are all concepts. So the concept of "mom" became more and more complicated over time.

Note that animals do not have this concept. While they do have short-term attachments to their parents, simply because their parents take short-term care of them, they do not have a *concept* of parent. Otherwise it would be a lifetime attachment. Needless to say, animals do not show any signs of understanding that their peers also have parents.

Consider the concept of an apple. What comes to your mind when you think of an apple? Red or green, round, sweet, tasty, a healthy fruit that grows on a tree, right? But the problem is that in nature there is no color, no taste, no health benefit, no fruit, and no trees. They all are products of our consciousness. All these notions are concepts we invented and as such they exist only in our imagination. The molecules that make up an apple have no color. The electromagnetic waves reflected from the surface of an apple have no color either. They only have wavelength and intensity. Our consciousness makes them red. The molecules also have no taste or smell. Once they hit our nose or tongue, they cause reactions that our consciousness *interprets* as sweet, aromatic, or tasty. Needless to say, the concept of "healthy" is as imaginary as it can get. Apples can be poisonous to a number of animals. What about the concept of "fruit"? This is also an abstraction that only exists in our minds as it is only a classification, and nothing more. Besides, there are all sorts of other apples in other languages and cultures. Thus, people who belong to those cultures or speak other languages associate many other notions with the word "apple". For example, in Hebrew, a potato is an earth apple, and an orange is a golden apple. In English we also have pineapples. And last but not least, there are no "trees" in nature. There is a collection of molecules (also colorless, by the way), assembled into a certain structure, which for some reason we decided to call a "tree" and classified it as a tree, so as not to confuse it with a bush.

What I meant to emphasize here is that even notions that seem very real such as "mother" or "apple" correspond to extremely sophisticated structures in our heads. These structures are related to reality only through very narrow subsets, such as your actual mother or an apple you just ate, while all the rest is as fictitious as money or corporations that have no connection to the real world at all, in the sense that they did not exist before we invented them.

Words denoting verbs, actions, are also concepts. What comes to mind when you hear the word "sit"? A meme, an image of a person seated in a chair? Not only that. It is also a dog performing the "sit" command, a rider *seated* in a saddle, a person *seated* on the ground. In other words it is any action associated with lowering the rear to a hard surface. Moreover, these are not all. Remember a bird *sitting* in a tree? It does not lower anything anywhere, it actually stands. And what about hurt feelings that *sit* deep in the heart? They also do not quite *sit*. Thus "sit" is an abstraction that covers a much wider range of actions and states than any of its values.

Thus, every word, phrase, logical construct, everything, absolutely everything that our consciousness operates with and consists of corresponds to some concept.

I hope by now you understand what a concept is. To those who had already figured it out, I apologize for such a long and tiresome explanation. My frustration was sparked by a journal referee who did not understand, or pretended not to understand. I agree that it takes some mental effort to realize that concepts are everything that the human mind operates with. But I think that it is quite doable for someone with an average IQ. Apparently, having false knowledge of the subject prevents such understanding.

It is clear that when I say that concepts exist in consciousness or are part of consciousness, long before I define consciousness itself, I am misleading you. It would be fair to ask, where exactly do concepts live, what physically or biologically do they correspond to? The answer is that each concept is a unique state (a sensory image) of the nervous system (the brain). Thus, each concept is hardwired into the brain with the help of feelings/sensations. Each concept corresponds to its own set of sensations (sensory image, state, profile). To learn, to acquire a new concept, we must train ourselves to *feel* it, and at the same time imagine what it means. In other words, we associate an image with sensations. Below it will become clear that every image *is* a set of sensations, and that this association is the same thing as seeing.

We learn concepts all our lives. We acquire new ones, change, expand, and deepen the old ones. Sometimes we get rid of obviously false or outdated ones. Many concepts have disappeared because they contradict science. Many were rejected as inadequate.

It's not easy to learn a new concept. To do this, you need to train your nervous system, connect the newly acquired concept with others learned earlier, and insert it into an existing system – that is, into consciousness.

Our speech is made up of concepts. Being an abstraction, each concept has a different meaning for different consciousnesses, depending on the experience and the whole history of development of this particular consciousness. Sometimes the difference is very small and the conversation partners understand each other well, but sometimes the difference is so wide that the listener does not understand the speaker at all, or understands something completely different to what the speaker assumes. This applies not only to such complex concepts as "I think", where the speaker, if properly interrogated, would have to admit that he does not understand the meaning of what was said, but also to seemingly very simple, everyday things. For example, when a mother tells her son to go and buy potatoes, he knows exactly what she means. This is because they have been living together for a long time, this is not the first time he has done it, and all ambiguities have long since vanished. Or perhaps where they live, there is only one place that sells only one type of potato and at one fixed price. As a result, ambiguities are not possible in principle. Now imagine that the same request was made to you by a person whom you barely know, in a place unfamiliar to you. You immediately have dozens of questions. Where should you buy them, at the farmers market or in a grocery store? If a grocery store, then which one, and what kind of potatoes, of what type, color, size etc, and at what price. It would also be nice to be sure that "potatoes" meant exactly what you thought, and not sweet potatoes, yams, or similar. If you do not know all these small things and decide them on your own, there is a reasonable chance that the other person may not be happy.

In this regard, I recall a real case when a client asked a programmer, who was my employee, to fix a program for printing documents. The program had stopped working after a system update. My programmer quickly fixed everything, but instead of the praise he expected, he received a reprimand and had to redo it. It turned out that instead of the beautiful color output that the programmer had made, the client wanted everything to be "like before" – black and white, with an outdated (according to the programmer) old font. The programmer was furious when he told me this. After all, he had done everything possible to improve the program (as he understood it), and yet the client was dissatisfied.

I tried to explain to the programmer that the client is always right, that what is "best" is decided by the one who pays, i.e. the client, and that before doing anything it is necessary to clarify with the client all the details of what exactly they want. Unfortunately, my appeals fell on deaf ears. The story has repeated itself over and over again. The programmer kept doing what he himself considered to be best, the clients were unhappy, and I had to fire this programmer.

The point I want to make is that what we say may have no meaning for the listener, or may have a meaning altogether different from what we intended. This is because the concepts that make up our speech are completely different for all of us, regardless of the fact that they sound and read exactly alike.

But let's return to information.

6. A definition of information

The previous section explained that information is produced by a human's consciousness (or in the consciousness of a human), and that its production requires concepts. By definition, the presence of an apparatus of concepts in consciousness means the presence of abstract thinking.

This is the ability to separate the property from the object and consider the property as a separate entity. Associative thinking follows abstract thinking: it is the reverse operation that takes place when an index, which is a concept, identifies a completely different phenomenon that shares the same concept.

By contrast, for animals, the received signal only has one meaning: the reaction that it evokes. Because animals are trainable, this reaction can improve over time. But it is never conceptualized. Animals do not systematize signals, they do not have concepts and do not analyze the "meaning" of the signal as we do when we identify various concepts. For example, a bird – in common with all other living beings apart from humans – has no concepts, including the concept of color. Simply reacting to color is not the same as having the concept of color. Supposing, a bird that ate red fruit suddenly started trying to eat everything red, then we might assume that it had learned the concept of red by wrongly associating it with everything edible. But this sort of thing never happens. If the reaction of an animal to a signal can be called information, then it is the most primitive particular case, so to speak, of an "information scalar", i.e. information that has only one component – the reaction to the received signal. For information to appear in the mind of the receiver of a signal, they must have a set of concepts necessary to produce this specific information. In turn, a concept places the receiver in a state of waiting for a signal. This explains why the absence of a signal, or the violation of a previously learned pattern, is an "informative" event.

I am not yet aware of any receiver of signals that has the above properties, other than human consciousness. I see no reason why these properties cannot be built into a machine. However, when this has been done, no one will be able to distinguish the consciousness of a machine from the consciousness of a human. In the meantime, we will assume that this is only about humans.

Thus, information is produced by human consciousness, and only by it. As it learns, it creates perceptual templates according to its set of concepts. We may consider these templates as questionnaires, or surveys, that consciousness fills out by scanning its sensors, and constantly updates the templates based on the results. We all know perfectly well that people of different political views produce completely different information from the same messages. This is because their questionnaires have completely different concepts and therefore make completely different requests when analyzing a message.

But what actually is the information that is produced by human consciousness as a reaction to the excitation caused by the received signal? Information is the result of the **analysis** of the conceptual composition of the response, and the analysis itself consists of decomposing the signal into conceptual components. Thus, **information is a conceptogram of a signal**, i.e. a set of concepts identified in the signal and the specific values assigned to them. Mathematically, information is a vector of concepts.

Consciousness decomposes the signal (or more precisely, the response) into all possible concepts (and their values), just as a medical laboratory analyzes your blood sample and provides you with its results, such as the level of glucose, the level of red blood cells, the level of cholesterol, etc. At the same time, the rest of the body perceives these results through tabulated sensations (each concept has its own sensation). I believe that all this is also accompanied by visual images.

In everyday language, **information is the meaning (interpretation by human consciousness) of the received signal.**

But at the same time, it is necessary to understand that 1) it is not at all about the "received" signal, and not always about the response to the received signal (which can be completely arbitrary, and even absent), but about the excitation of neural networks that is produced by our brain, both arbitrarily and as a result of "receiving" an external signal; 2) "meaning" is a set of concepts with the values assigned to them in the case of the signal received; 3) Interpreting is the process of decomposing the signal into concepts and assigning values. At the same time, it is clear that in another's consciousness, or in our own consciousness at another time, the result of interpretation, i.e. meaning, can be different, or even completely absent.

Imagine how many concepts we have not yet come up with, and therefore how much new information people of the future will find in the same signal? Most likely they will assume that we did not know anything, or were wrong about everything.

Some call this constant chain reaction of signal exchange in a brain a flux of consciousness or a flux of information. Each signal is interpreted as information. As a result, consciousness has consolidated predetermined patterns that evolved in the course of its development, experience, and education. We know from our own experience that our perception of the same signal changes not only during life, but even during a day, depending on our state of consciousness.

Information, as well as each individual concept, is identified by the totality of sensations it evokes. There is also a functional cluster which our body associates with the concept "I" and deceives itself into believing that "I" is thinking and is a controller of all the various processes. It is clear that each of our concepts, each word, action, logical operator or logical construct, has its own reflection in some cluster (and most likely not in one but in many competing functional clusters) – a template with its tabulated sensations that it sends to other clusters, including "I", unless, of course, the "I" is asleep or resting.

7. Consciousness

Thus, human consciousness is a generator and a processor of information. This function is an extension of the ancient animal functions of the brain, namely observation, intelligence, and response to change. The task of consciousness is to observe, to see and react, or in other words to give the rest of the body a reaction to what is seen. The eyes are our visual sensors, and the brain is our internal display organ. All consciousness is visual, including information and thoughts. Any information, any thought is an image, a picture, a clip. Even sound information is instantly translated into a picture. Pictures are constantly compared with existing templates in order to detect changes. However, while an animal's consciousness is limited to this, a human's consciousness has additional vision, which could have been brought about by an excess of biomaterial, namely, "free" ("jobless") neurons. Perhaps as a result of a lack of assigned work, these neurons began observing the work of others, "completing" the picture of the "seen" by guessing, and creating an entirely new thing that was not in the original image, thus producing information and conceptualizing.

To display a picture of the "seen" and give the body a reaction to the "seen" is the whole purpose of this constant exchange of signals between neurons. I put "seen" in quotation marks because it does not matter to the brain where the signal originally came from – whether from external sensors or generated by the brain itself. While the brain is alive, this work never stops, neither in a dream nor in a coma.

The electrochemical processes themselves, the excitation of various parts of the brain, are observable and well studied by researchers involved in mapping the brain [2]. But the pictures that correspond to the sensations associated with these signals cannot be seen by an outside observer. It is impossible to see the sensations that a particular signal causes in a given individual's consciousness. These pictures-sensations depend on the whole history, experience, education of a person. Therefore, it is not surprising that scientists who map the brain and study its work, while having made good progress, nevertheless attribute some sort of mystery to consciousness, as to something immaterial, detached from the brain itself, or if it is related to it, then in some superficial way. As one neurosurgeon famously said: "I've opened a brain many times but never seen consciousness in it."

Cinema is a good illustration of how it works. When we watch a movie in the cinema, we know very well that it is just a flat screen that reflects lights coming from the projector. So, what causes us to fear that train apparently coming at us, or gasp as if we really were jumping from that plane, or see and feel ourselves as if we were really in the middle of that epic battle, or love scene? What on that flat screen causes us to laugh, cry, scream, hate, as if it were real? The sequence of images that imitate real-life motion triggers exactly the same sensations that would be triggered by the real events. The consciousness then "completes the picture" by triggering all the sensations that would correspond to the real events, which are in turn reflected on the brain's internal screen as a result. Thus, the cinema screen creates an illusion in our consciousness that we are witnessing or participating in real events, because it reproduces exactly the same sensations that the real events would create. Consciousness is this never-ending movie, where "I" is associated with the spectator. In addition, it is a multi-screen performance with hundreds of sub-screens that run various fluxes of consciousness, or channels, such as memories, history, news, daily events, business, science and so on.

Only you know your feelings and sensations. And even when something hurts, your doctor doesn't "see" the pain. That is why she asks you to describe it. All your doctor can see, using any of the most advanced tools, is whether or where you have inflammation and what parameters are abnormal. But it is impossible to unequivocally determine whether you are in pain or not, as only you can know this. Similarly, it is impossible to see another person's consciousness as a whole, because their feelings are invisible to everyone else.

Moreover, you do not feel all of your own consciousness! This is what I was talking about at the beginning when I asked you HOW you think, or when I described situations where someone else "thinks". Not all of our consciousness is open to us.

Thus, **consciousness is the sensation of the brain's work**.

Sometimes consciousness is understood as the totality of the work of the brain, and not just what is felt. But this is just a question of convention, and is of no interest to us. Sometimes the work of the brain that is not felt is referred to as sub-consciousness.

The definition of consciousness as a feeling of the brain's work is perfectly consistent with everyday parlance. For example, "to be conscious" and "to feel" are practically synonymous.

But the question remains: who feels this work? Of course, the rest of the body. In the case of humans, there is a non-existent someone whom we call "I", to whom we mistakenly attribute various functions of the brain. "I" is a concept, just like any other. But in some ways, it's not quite the same.

8. I

Consciousness is a completely real, material sensation. It is no less real than the feeling of pain or fatigue. While we can feel pain and fatigue very early in life, the subtler components of consciousness, such as disappointment, sadness, compassion, love, resentment, envy, and the thought process, are felt much later in life. These sensations emerge and develop as they are learned at later stages of personal development. The human "I", which in itself is also a collection of sensations that is acquired long after birth, develops throughout life. It constantly expands, changes, and adds various sensations. The acquisition of a novel sensation consists in assigning functional clusters of neurons to tasks such as the identification and tracking of certain other sensations, reaction to their changes, communication with other clusters and the organism as a whole. All this is well known to brain researchers, who map the brain, and to biochemists and pharmacists, who are well aware of which sensations correspond to which electrochemical processes and even produce mood-altering pills based on this.

Thoughts, logical constructs, and all the work of the brain that goes into producing and processing information are felt by the body just the same as any other subtle sensations. We learn these sensations just as we learn all others. Of course, such a simple thought as "I'm hungry" comes to us much easier, gets tabulated at earlier stages of the formation of self-consciousness, and is felt much more strongly than a complex mathematical or logical construction. However, even this simple thought is not mastered by everyone. There are people who do not experience hunger at all, even when their bodies are dying from lack of food. Their consciousness either did not create this cluster, or blocked its work at some point in their lives. Not to mention the fact that complex logical constructions may never emerge if the appropriate training has not taken place.

When we "think", we attribute this work to our non-existent "I". It seems to us that we are consciously forcing our brain to work in the direction "we" have chosen. But in reality we are confusing cause with effect. The brain is the one who thinks. It "thinks" about everything at the same time. The functional cluster that we attribute to "I" creates a feeling of ownership by observing, tracking, and following a particular direction of the brain's thoughts. It claims ownership of this work and attributes it to itself. To make "yourself" think, you need to activate the corresponding templates of "thinking", i.e. functional clusters responsible for it. These clusters are monitored by a "controller" that produces emotions (based on a previously tabulated template) indicating if it is "this" thought or not, and "corrects" it if it isn't (by acquiring corresponding templates). In general, although everything is done strictly according to templates, it seems to us that it is "we" ("I") who "think". Only, for some reason, our "I" has no idea HOW it does this. All it does is follow, track, and observe the work of the brain, and assign the results to itself. In order to do this, "I" needs to be in a state of alertness and focus on the results, in order to quickly react to them and to the results of their evaluation by other clusters according to their templates.

It is like a theater without a director, without a script, in which there are only actors and spectators. The actors do something on the stage. The audience has only one binary function – to react, either by applauding what is done on the stage, i.e. approving, or booing, i.e. disapproving. The actors adjust what they are doing in order to get more applause and fewer boos. As a result, depending on the talent of the actors and the taste of the audience, there is some kind of action. Sometimes it works out well, sometimes it doesn't. An outsider would never guess that there is simply no director or script behind all of this. Now let's imagine that the audience has its own spectators, who express their satisfaction or dissatisfaction with the way the first audience reacts to the play. A third level of spectators watches them, and so on. And now let's go even further and turn spectators into actors too, in other words we will not differentiate between them. Everything will work exactly the same. A certain self-organization arises, a certain action, behind which there is no director, which in our case is "I". He or she simply does not exist. There is only an illusion that they do. Of course, we can denote by the word "I" the non-existent director-producer, which effectively is all of this together – all the actors-spectators (neurons), their interactions and results – but the essence of the situation does not change. "I" is a kind of inner god who, just like an outer god, is a representation of everything unknown. "I" is a hidden "producer" of chaos with only the appearance of intelligent design.

It is very natural for our consciousness, which must see everything, know everything, and have an explanation for everything, to create concepts of I, God, and similar. This is because they create the feeling of the world being explained. And that's the only thing that matters – the feeling of understanding, rather than understanding per se.

Like any other concept, "I" is hardwired into our brains as a specific sensory image. It is constantly expanding, changing, updating, and attributing new functions to itself or getting rid of old ones. The "I" of today may be quite different from that of yesterday. I suspect that it may even change throughout the day. And sometimes there are several of them. My "I"s even argue with each other. I think yours do too.

Actually "I" is an imposter. Being a regular concept, a functional cluster, like many others, it took over the function of a leader. This is because somebody had to, in order to satisfy our desire to feel complete. "I" himself has no other function than watching others work, and sleeping a third of the time. It is more correct to say not that "he took over", but "we attribute to him" these functions, despite the fact that most of time he rests, dreams, contemplates, and reflects. An extremely lazy creature.

Our "I" is a kind of little prince who does not know anything, but noticed that as soon as he thinks that he is hungry, all kinds of goodies appear on the table. How can one not come to the conclusion that all this appeared because of his thoughts? But ask him (like I asked you in the beginning) HOW exactly his thought was transformed into what is on the table, and he has no idea. He does not know, and does not need to know how and who grows wheat, grapes, raises the livestock, how they then make bread, wine, prepare meat, how and who brings all of it to the kitchen, cooks it, and serves it to him.

We also convinced ourselves that "I" thinks, since he associates the function of verbalizing the overheard thoughts with himself. What he calls a "thought" is not an image flashed from the sub-conscious, but its verbal expression. He considers "thinking" being not browsing through thought-images, but following an internal conversation that goes on in the consciousness. "I" tries to steal this conversation by attributing it to itself, though not always successfully (remember the "elusive thoughts"). I categorically claim that "I" is a liar, a plagiarist and a usurper. He doesn't do anything at all.

The entire function of "I" is to observe the work of consciousness and appropriate the results. And the body perceives it as thinking. As a result, thinking still happens but in fact it is both more primitive and more complicated than it might seem.

9. On material vs. immaterial

Consider the transformation of an information vector. As mentioned above, information can be thought of as a vector of concepts. Let N be a number of concepts a_i (i = 1,2,...,N) in some particular consciousness. Denoting the input vector of information as $A = (a_1, a_2, ... a_N)$, and the resulting output vector as $B = (b_1, b_2, ...b_N)$, we can represent the converting operator in the form of a matrix $S = \{ s_{ij} \}$:

$$S \cdot A = B \tag{1}$$

or, which is the same, $s_{ij} a_i = b_j$.

Here we have connected the vector of a perfectly material signal A with the vector of an equally material response B (both are materially felt by the body), by a completely immaterial field S that we have just invented. And now it seems to us that S is something real, though immaterial (and in a sense it is). But we just came up with it on our own. Actually there is no such thing as S. The only "real" entities are A and B. Though they are concepts too. All we have in connection with S is an image of a formula (1) and a concept S in our brain. But this is a fiction. Our entire consciousness is a similar fiction since it consists of the field of such fictions. I think it's time to realize that there is nothing mystical about consciousness. The whole reason for the perception of it as a mystery was in using it without bothering to study the process. And since, after acquiring abstract thinking, it became necessary for us to have an explanation for everything in the world, we, without really bothering much (as we often do in many other things), were satisfied with all sorts of mystical interpretations that entered our language so strongly that even having understood the real meaning of consciousness, it is difficult for us to formulate it.

Consciousness is just one of many concepts, fictions,

illusions that we operate with quite successfully. In physics, we relate one fiction to another, and everything works well. We live in a world of events, in which there is no continuous time, let alone space, no "forces of nature", no mass, charge, spin, or color of quarks. These are all fictions invented by us (see the final chapter 13). And yet, we sometimes manage to connect one fiction to another and build a ship, a car, an airplane, an iPhone, an atomic bomb – all of them very real things.

We consider something to be material if it can be touched, seen, tasted or smelled, or at least one can imagine that what is seen (even through a microscope) has such properties. The immaterial is something that cannot be seen, touched, tasted or smelled, even in theory. The latter include visual images (as opposed to their sources), language, mathematical formulas, everything related to consciousness (or the products of consciousness), and consciousness itself.

But, firstly, as we have already understood, all these sensations of material entities ultimately come to us through our consciousness, and secondly, both material and immaterial have a completely material realization in our brain in the form of sensory images (representations or profiles).

Thus, due to concepts, there is a certain duality that emerges: on the one hand, we have a sensory (physical, biochemical) image (a representation), while on the other, we have its seemingly immaterial projection in consciousness. By the way, this is the essence of the dispute between Dennett and Chalmers. Translating into the language of this text, Dennett claims that this projection is an illusion, that there is nothing beyond sensations, and that everything else is mistakenly perceived by us as a separate entity. Chalmers, by contrast, says that there is a projection, in some dimension unknown to us, which we cannot feel in any way except through consciousness. I also think that this is the line dividing those who believe in God and materialists.

So is there a projection? Where is the screen? Or in other words, who paints an apple red, and with what colors, if red does not exist in principle?

This naturally leads us to the next chapter.

10. Vision is consciousness, and consciousness is vision

Having emerged as a device for displaying the "external" picture, consciousness began to work on the visible picture by completing it, filling in missing details, guessing and inventing everything, including what is not there. In other words, plain lying. I am far from thinking that there is a projector with a screen in the brain, but functionally, through our sensations, this is exactly what happens. External optical signals activate the same sensors as signals produced in the brain itself.

Most people probably guess that consciousness and vision are somehow connected. It is not a coincidence that both disappear when we fall asleep and return together when we wake up. We also know that under the influence of alcohol or other substances that affect consciousness, vision is drastically impaired. In a state of fear or extreme affect, vision becomes tunnel-like – we cease to perceive the "side" picture. Professional illusionists have developed a lot of tricks to make the viewer "see" something that is not really there. From all of this, one might think that consciousness and vision are somehow related to each other, but are not necessarily one and the same thing. I claim that consciousness and vision (perception) are simply one and the same.

The reason we use two different words to refer to this entity is that, thanks to our deceitful language, by "vision" we mean something that we do by means of the eyes and by "consciousness" something we do by means of the brain. We tend to think that what we do with our eyes is simple and understandable, whereas what we do with our brains is complex and mysterious. Consider what our eyes actually do. They, together with all the electrochemical channels attached to them, are just optical sensors that respond to electromagnetic waves in a very narrow range, from about

380 to 700 nm, and send the corresponding signals to the brain through complex processes and neurons. But we see not with the eyes but with the brain (consciousness), which is the real organ of vision.

Ask a physicist why we see grass as green and sky as blue. He will be happy to tell you about wavelengths, about the reflection of light, about how both change depending on the properties of the reflecting material. He will have nothing to say about how all that turns into a color that we actually see. An optometrist or ophthalmologist will tell you how the signals from the retina get transmitted to the brain, but he too would be unable to answer this question and would have to refer you to the mysteries of consciousness.

In fact, electromagnetic waves are not red, blue or green. The response they trigger in the eye is transmitted to the brain by means of electrical and chemical signals that carry neither color nor geometric shapes. It is consciousness that "colors" the "seen". The leaves on the tree are also colorless. They simply change their molecular structure depending on the weather and seasons. Their reflective properties change accordingly. Moreover, the reflection differs according to the amount of daylight. So on a sunny summer day they reflect mainly 550 nm (what we perceive as green), on an autumn day 570–580 nm (yellow) or 620–700 nm (all variations of red), and at night there is nothing to reflect which is interpreted by brain as the color black.

Similarly, the sky is not blue and the grass is not green. There are no "colors" in nature at all. They exist only in our minds. Also, there are no forms because they too are a result of the work of the brain.

In nature, there are also no beautiful melodies and no pleasant voices. Just as our eyes respond to electromagnetic waves, our ears respond to acoustic waves, i.e. waves of air density in the range from 1.7 cm to 17 m. Beyond this range we cannot detect anything at all. Acoustic waves are not pleasant or unpleasant sounds or melodies. Just like electromagnetic waves, they have only two characteristics – amplitude and frequency. It is our brain that reacts to a set of these air-density fluctuations and interprets them as Beethoven's *Moonlight Sonata* (which, by the way, for many people is nothing more than a terrible noise), or Bach's *Toccata and Fugue*, and tells us that they are beautiful (if this is how we were educated). Animals do not perceive this beauty, just as they are not inspired by the views of mountains, forests, and the sea. They don't have a concept of beauty. (Incidentally, if not for the air, we wouldn't hear anything at all. Unlike us, mosquitoes can hear ultrasound. We could try transposing the *Moonlight Sonata* into the mosquito's range, just a few octaves higher, and see how they react, although we wouldn't hear a thing. As regards the pleasantness or unpleasantness of sound waves, I am oversimplifying a bit here, as there are still consonances and dissonances. There is a physical reason why some combinations of frequencies form harmonies while others don't. Yet, this is beyond the point of this discussion, which is the subjective perception of sounds by our consciousness.

Equally, there is no taste or smell of food, or smell of perfume, or anything else that we enjoy or, on the contrary, feel disgusted by. The natural world is a world of molecules that have neither taste nor smell. Through the nose and tongue, they trigger electrochemical reactions in the corresponding receptors. These signals then travel via nerves to the brain, where our consciousness creates the sensations of smell and taste. It is our brain, or rather consciousness, that tells us that strawberries are sweet and aloe is bitter, that fresh meat is tasty and edible, while rotten meat is not.

The same is true for tactile sensations, which are actually the result of our skin receptors interacting with outside material. The result of the interaction is transmitted to the brain and interpreted by consciousness as "smooth", "delicate", "rough" or "coarse".

Thus, the brain (the nervous system as a whole) converts received external signals into visual (perceptual) images. Therefore, what we call consciousness is actually vision in the most general sense of the word.

Therefore, we can conclude that consciousness is just as material or immaterial as vision. If what we "see" is in fact an illusion, and nothing more than a sensation associated with concepts, then Dennett is right: there is nothing immaterial. If, on the contrary, there is a projection beyond a material carrier (brain, nervous system) into some immaterial space where we see all of this, then Chalmers is correct. The theater illustration above provides answers. While the imaginary producer of the play ("I") is absent and in this sense is an illusion,4 the sensation of it is not. As a result, the "hard problem"5 is resolved since it is a very real phenomenon encoded within the brain and nervous system.

But let us recall that human consciousness consists of concepts. The question then arises: if consciousness is vision, what is the role of concepts in the formation of visual images? Does the absence of concepts mean the absence of vision? The answer is no, it doesn't. There can be no doubt that the absence of concepts in the mind of an animal, as well as in the mind of an infant, does not mean that they do not have vision (including hearing, smell, taste, and tactile senses – all of which have the same purpose, which is to see, to imagine, to understand, what is there). Then what do they see? To answer this question, try to remember yourself in infancy and try to recall at least some kind of visual memory. I have no doubt that you cannot recall any pictures you saw before you were a certain age, just as I don't. Depending on the person, it could be 3 years old, 5 years old, or even older. In other words, before the formation of the first concepts, especially the concept of "I", we did not have any visual memories. This does not mean that we did not see. It means that we saw subconsciously, i.e. just as animals do and our human ancestors did. This subconscious vision is intended only for an instant reaction to what is seen and is not burdened with any further processing, interpretation, and memorization. Infants, animals, and primitive humans cannot perceive two-dimensional images, photographs, paintings, or films. And even the concepts of color, well known to primitive humans, judging by their bright costumes and cave paintings, do not appear immediately in children. You have probably heard children ask each other or adults what their "favorite" color is. This means that they have already learned to distinguish and perceive colors, but do not yet know what to do with them.

I am always amazed to observe the children of tourists visiting national parks such as the Grand Canyon, Glacier Park, Yellowstone, and Yosemite with their breathtaking, spectacular views. While the adults watch speechlessly, stunned by the landscapes, their children – just like their beloved dog (I apologize for the comparison) – run around, not even slightly impressed with the views, barely noticing them. They don't see the beauty all around them. Children cannot see this yet, while dogs will never be able to see it. They do not have the concepts of beauty, harmony, and similar attributes of adult humans' consciousness-vision.

Surprisingly, our language, which deceives us about almost everything, sometimes adequately describes some processes in the mind. Expressions such as "see the difference", "look for the cause", "I see" (in other words, I understand), political "views", and so on, use the terminology of optical vision, although they refer to consciousness.

It is also understandable why it is so easy to deceive us. We are constantly searching for explanations of what we see, for the "complete picture". In the absence of our own explanation, we will happily buy whatever is offered to fill the void. We must know everything, understand everything and explain everything. We are better off with a false model, information, or concept than not having them. This is the human tendency that magicians, illusionists, swindlers, politicians, kings, dictators, priests, and journalists actively exploit.

Some people may feel upset to learn that their brains are so "primitive". On the contrary, I find it amazing how out of simple neurons, whose only function is to exchange signals with their neighbors and react to the signals they receive, such a complex functional structure emerged. This functional structure not only arose, but actively teaches itself, programs, thinks, feels, reflects, and strives to understand the Universe. I emphasize the word *functional* here. While the number of neurons in the brain is finite, the number of functions is infinite. Moreover, it is clear that the same neuron can be included in an infinite number of functional clusters, and its small function can be transferred to another.

Other people may be frustrated by the inevitable conclusion that consciousness, thinking, "I", and other concepts do not correspond to anything more tangible than sensations. I see no reason for frustration. As Harari correctly noted, almost everything that surrounds us – money, states, corporations, banks, etc. – are illusions that exist only for as long as we believe in them [12]. So what? That doesn't make them any less effective. Thanks to these illusions, we have a modern civilization. We have the opportunity to develop, improve our lives, and engage in science. Even though our thinking is an illusion to a certain extent, the planes, satellites, ships, cars, artificial materials, atomic energy and atomic weapons created by science are by no means illusions.

We should be grateful to our illusionary consciousness for the colorful world around us, for the opportunity to enjoy music, food, wine and much more, despite the fact that the real world has no colors, no music, no taste or smell. Our consciousness makes our lives worth living.

We could stop here since we have answered what consciousness, information, and "I" are. Nevertheless, there are some remaining questions which, although they may seem secondary, from a practical point of view are perhaps no less significant and require more detailed consideration. Among them are the origins of reason, logic, and logical constructions.

Consciousness exists in all living beings that have a nervous system, since consciousness is a sensation that arises from it. But what distinguishes a human from an animal? The answer is the presence of an apparatus (or module) of concepts that constitutes human consciousness, or what we call reason. Reason is responsible for logic, abstract thinking, speech, the analysis and systematization of observations, the generation of information, and other things that belong to the human domain only.

Below we will speculate on how the reason that made us *Homo falsus* (a.k.a. *Homo sapiens*) could have emerged in our animal ancestor.

11. The origin of reason

Why is it so difficult for us to learn, perceive, and create new knowledge? It is because all of this requires the creation or acquisition of new concepts. Even acquiring the simple concepts of "mom" and "dad" is not easy for a child. Developing a new concept is an energy-consuming process. As such, in physics terms, it must be a phase transition.

At the biological level, i.e. at the level of neural clusters, the creation of a new concept entails the destruction of an existing adjacency matrix of the human connectome [13] and the establishment of new connections. This must be an extremely painful operation, even though we do not feel this pain as regular pain. Moreover, the more radical the new concept is, the more painful and energy-consuming is its creation.

This is why many students prefer mindless memorization to the acquisition of conceptual and systematic knowledge. In addition, many people stop learning altogether at some point in their lives. Obviously, there is a huge difference between learning an everyday concept and learning a scientific one. On the other hand, despite the pain and significant energy consumption needed initially for the creation of a concept, acquiring concepts is energetically advantageous over the long term. This holds true for an individual and for the species as a whole. This is because concepts facilitate the acquisition and transfer of knowledge, and allow prediction and forecasting.

My hypothesis is that reason, which is the module of concepts, emerged in the mind of our animal ancestors through a phase transition. As a result, their consciousness transitioned from a state of the plain memorization and accumulation of facts (the accumulative phase) to the phase of conceptualization, i.e. the generation of concepts (models, theories, systematic knowledge).

It therefore provided the ability to analyze, systemize, generalize, and predict.

Just as humans created concepts, concepts created humans.

Purpose of life

Why do we constantly seek out information? What was it about the transition from a simple accumulation of facts to the creation of conceptual knowledge that led to the historic, qualitative leap from animal to human? To answer these questions, we need to think about nothing less than the purpose or meaning of life. Without a clear understanding of what it is, we cannot answer this question.

If we step down for a moment from the pedestal on which we have placed ourselves, if we give up for a moment the arrogant confidence in our uniqueness and importance – the belief that the whole world exists only for us, with us being its center, meaning and purpose – and honestly analyze our behavior compared with that of animals, we will find that we are not all that different from them.

This little difference is our human consciousness, which:

- most likely emerged by accident,

- may well emerge in other animals,

- will soon emerge in robots, with our help.

Unfortunately, we must make assumptions in these notes that we would not have to make if the human biological sciences were at a more advanced level with regard to consciousness. For example, the assumption about the possibility of the emergence of human-type consciousness in animals, as well as the very mechanism of its emergence in us, will obviously sooner or later be resolved by biologists.

If we could produce clones of our ancestors – brought back to life like the fictional dinosaurs in *Jurassic Park* – we could, by comparing them before and after the transition, pinpoint the time period when the transition occurred. If the above is true, I am ready to offer a completely unambiguous criterion of how to distinguish a human from an animal. This criterion is the ability to acquire concepts. Of course, if you are dealing with a creature that has not yet developed speech organs, you will not be able to teach him a spoken language. But most likely you will be able to teach him many other things that are impossible to teach an animal, maybe even mathematics. If anything, the difference should be huge and obvious.

Therefore, if we rein in our arrogance a bit and agree that we are nothing more than talking (and abstractly thinking) primates, then it will be easy to reach the conclusion that the purpose of our existence, just like for non-speaking primates, and every other animal is nothing more than **to spread our DNA**. In other words, and sadly, we are just a temporary shell for our (almost immortal) DNA – our "selfish genes", as the evolutionary biologist Richard Dawkins memorably called them [14]. This shell is no longer needed after we complete our mission of passing on our DNA to our children.

From our birth until our death, we and every other animal strive to do the same:

1)   Support and extend our own life (as carriers of DNA),

2)   Give birth to as many offspring as possible (i.e. transmit DNA),

3)   Provide conditions for the survival of these offspring (i.e.

the survival of DNA).

Obviously, I am talking about the biological purpose of life that humanity has been given by nature. Everyone is certainly free to think that the meaning of their life is to do good, to love, build a house, plant a tree, make a scientific discovery, or just enjoy life, but if you honestly examine all these "meanings", you will find that all of them aim to accomplish one or more of the above tasks.

DNA temporarily uses our bodies to multiply, passes itself to our offspring, and then discards us as a spent and no longer needed biomaterial. Although nature does not forbid us to enjoy our short existence along the way – to love, learn, and so forth – make no mistake, this is not the true "goal" of nature. It uses us only as temporary carriers of DNA: a temporary shell containing a biological portrait (or template) of a species.

Hunters for information

To accomplish all three of the above tasks, we, like all animals that have the appropriate sensors, need to constantly monitor the environment, detect changes, identify danger, food, and do everything else that is necessary to sustain life, identify a mating partner, and so on. From our human point of view, we would call it hunting for information, although, as we have already established, animals do not produce information, only we have this ability.

This hunt for information is necessary to accomplish the same three goals:

1)   For the survival of the DNA carrier, it is necessary to find energy, food, heat (or money to buy them). This requires *finding and identifying* the sources of these and methods for acquiring them, as well as observing and identifying

52

possible dangers to life and health.

2) For the transfer of DNA to offspring, it is necessary to *find, identify* and *acquire* a partner.

3) To provide conditions for the survival of offspring, and to a certain extent for the transfer of accumulated "knowledge" and skills, it is necessary to find the means of information accumulation and transfer.

All of us, including other animals, are hunters for "information". Whether we like it or not, our brain (as well as the brain of the worm) is constantly processing incoming signals and trying to find a solution to one or more of these three tasks. Even when it appears we are in a purely contemplative state, our brain constantly scans and evaluates incoming signals for their potential usefulness to obtain energy (or money), position (which eventually turns into the latter), a mate, or at least some benefit for our offspring. This scanning takes place in our sub-conscious, at the animal level. As our brain is doing it, we can think about some philosophical problem, cook food, read, watch news, or do nothing at all. One does not interfere with the other.

In addition, higher forms of life are equipped with a reward system in the form of endorphins, adrenaline and other drugs that are produced by our body when the hunt ends successfully, whether for the goal is food, money, a partner or "information" about them. Orgasm is familiar to almost everyone. This is a reward for the successful transfer of DNA. And it does not matter in the least that the goal of a particular individual may be orgasm itself, rather than the transfer of DNA. Nature is indifferent to this distinction, as it has provided us with this reward for the purpose of the survival of the species. The fact that some people use this reward purely as a "drug", rather than for its intended purpose, is irrelevant to nature[6].

Even at the mere sight or anticipation of a trophy, we experience euphoria. For instance, "love at first sight" is nothing more than the euphoria induced by the drug injected into the body upon recognizing the "right" partner, and the anticipation of a successful "hunt". Of course, the image of the "right" partner (there may be several) is encoded in our subconscious in the form of a template that includes appearance, smell, movements, gait, gestures, voice, and so on. Identification happens instantly, even before we have time to think about something with our upper (slow) cerebral cortex [15].

The well-known buck fever (the shivers that seize a novice hunter at the first shot at a deer) is nothing more than the result of an adrenaline rush. Sometimes the trembling is so strong that the hunter is not only unable to aim, but even to hold a gun in his hands. Forget about accurate shooting. Eventually, after a few unsuccessful attempts, the hunter learns to control his trembling so that it does not interfere with his shooting.

These are the most striking examples of hunting, in their original form. The hunt for "trophies" and for information about them.

We can also observe the hunt for "information" in the animal world. Animals are very curious. Hunters know that this is one of the factors (besides the search for food and a partner) that make an animal forget about fear and often entices it to come closer. Curiosity is the inherent instinct of seeking information.

In modern, civilized society, hunting has more subtle manifestations. The euphoria of a businessperson, financier, or

---

[6] I hope the reader understands my sarcasm in using terms such as the intentions or purposes of nature. Of course, evolution has neither intentions nor purpose.

trader following a successful transaction, or of a scientist who just made a discovery, or simply of a thinking person who suddenly understands something profound – these are all manifestations of our "hunting" instinct. I don't know the answer to the question of whether a tiger experiences euphoria when catching prey, or whether it is only hunger that compels it to give chase, but one way or another, nature has provided us and animals with all possible incentives for hunting, including hunting for "information".

What distinguishes humans from animals is the fact that, having been hunters for "information" in quotation marks, we learned to produce information without quotes, in the sense that we discussed above. This new skill immediately separated us from the animal world. We continue to excel at this. We have already reached the point where our desire to have a concept, a model, a theory, an explanation for everything in the world has become our biological instinct. This instinct is so strong that we, as a species, with the exception of rare individuals, prefer even a false theory to its absence.

Of course, the creation of concepts requires much more energy than simple mechanical memorization, but this is a one-time effort that pays off over time because, firstly, it facilitates and speeds up further learning, and secondly, it allows us to make predictions. Accordingly, it is much more profitable for us, both individually and as a species, to have conceptual, theoretical knowledge rather than random (phenomenological) knowledge.

Experience vs. intelligence

Here and below, unless otherwise stated, we discuss human consciousness, or reason, and use these terms interchangeably.

From our own experience, we know that the mere memorization of facts does not raise the level of our intelligence.

It only makes us more aware, at best more experienced, but in no way "smarter" or more intelligent. A new level of understanding, and consequently of intelligence and consciousness, is achieved only when we manage to generalize and systematize the available facts. In scientific terms, this means creating a model: a theory in which they fit. In plain language, it means converting experience (empirical facts) into intelligence, i.e. into information.

The conversion of experience into intelligence is the work of reason. Each such conversion corresponds to a simultaneous increase in intelligence.

It is worth noting that the accumulation of experience occurs **smoothly**, whereas intelligence, on the contrary, changes **by leaps**. The leap occurs when a new concept, a new model, emerges in consciousness. All the random "facts" then fall into place, like spins in the Ising model. The newly created model immediately makes it possible to predict. If the prediction turns out to be correct, it confirms the correctness of the model, or refutes it if it turns out to be wrong. This situation is familiar to every researcher, as well as to those schoolchildren and students who were lucky enough to receive a scientific education. This is where they acquired not just random facts but models that explain the facts. If they did not acquire such models, I would suggest that such an "education" is a sham. It is just the mindless accumulation of facts. Such "knowledge" can be replaced by a Google search, which many students happily do. However, this kind of "knowledge" does not increase intelligence. Without understanding and remembering, you would have to ask Google how much 2x2 is every time you need this information. It may be possible to survive like this, and even prosper, but this has nothing to do with intelligence.

Thus, human intelligence develops in leaps and bounds, from one generalization to another. Some manage to do this all their lives, whereas others do not progress beyond high school.

A striking example of the advantage of scientific education over experience (without minimizing the value of the latter) can be seen in numerous job descriptions that appear regularly in many job bulletins. For example, a job description for a programmer position will routinely contain something like this: "10 years of experience in XYZ (an imaginary programming language), OR 3 years of experience AND a master's degree, OR 1 year of experience AND a doctorate degree." These are not arbitrary requirements, but the result of the experience of an employer who has learned that a scientific degree is equivalent to many years of experience. Moreover, when an employer cannot find an experienced programmer and the job still needs to be done, they are more likely to hire someone who holds a scientific degree, in the belief that he or she will learn the required skill much faster than someone who does not possess a degree but has experience in a similar field.

We can understand intuitively the jump that occurs when facts fall into place to create a new model. But is it possible to describe it formally? Any theoretical physicist knows that a jump is a phase transition. Although I have little doubt that the emergence of a concept, a model in the consciousness of an individual is a phase transition, the formal mathematical description of this phenomenon is beyond the scope of this book.

Similar to an individual transition, the acquisition of a concept by a group of people, or even by the whole of humanity, is also a phase transition, but now of a collective consciousness. In order for the phase transition of one individual to become the phase transition of the entire community, it is necessary that the same phase transition occurs in the minds of all members of the community. A collective phase transition requires considerable intellectual work by other members of society, although to a lesser degree than by the "discoverer". The work of the secondary members (colleagues who have read the article of the discoverer, for example, then students and schoolchildren if the result has become part of the curriculum) is facilitated by the fact that they know the result in advance. They still have to do some intellectual work in order to comprehend the result. Consequently, the result becomes their own discovery. Sometimes this leads to unintentional scientific theft, when as a result of such work, a person forgets that he received the idea from someone else, and did not invent it on his own.

The emergence of reason

From here, extrapolating into the distant past, it would be logical to assume that the **emergence of the first concept, i.e. the very emergence of reason**, was the same phase transition that we regularly make in our own brain (for those who do).

I think that sooner or later science will be able to unambiguously establish what served as the biological prerequisite for such a transition. I would venture to suggest that the prerequisite was a surplus of neurons, neural networks, and functional clusters. Those that were not assigned specific functions for active participation in the general process of observation began acquiring their own functions. Metaphorically speaking, by "thinking up", "guessing", and "completing" the details that are missing in the picture observed by their peers, looking for something that isn't there, or even blatantly lying without malicious intent, they are able to "justify" their presence to themselves and to others.

But unlike the biological prerequisites for such a transition, I do not think that we will ever be able to unambiguously establish what exactly served as the trigger and the motivation. In the language of the science of phase transitions, the external field is motivation, which can be any strong emotion, and the trigger is a random event. It is equally unlikely that we will be able to unequivocally establish what exactly was the first concept that led to the emergence of the whole apparatus of concepts. Therefore, there is an endless opportunity for speculations, conjectures, and hypotheses, which I immediately and gladly indulge in.

Hypothesis 1. Evolutionary emergence of reason

As already mentioned, our brain, being a visual organ, is constantly working on generating pictures of the seen. It is easy to understand that the desire to recognize what is observed leads to conjecture, to adding details to the picture, to completing it. You saw ears flicker in the grass, and your imagination (and fear) immediately drew one of the familiar concepts, such as a tiger or a wolf. Having seen just a few details, you have filled in some of the missing ones. If the guess was wrong, then the added details are also wrong. When danger is involved, there is no time to confirm a guess. It is always better to retreat, even if you are wrong about what you just saw. This is how concepts (threat, danger) and abstract thinking (adding a picture, inventing missing details) emerge. Additionally, it is necessary to immediately tell your relatives or peers what you just observed (and maybe tell them an unintentional lie) in order to warn them about the danger. This is how the need for language emerges. The more details are seen, the more reliable the guess becomes.

Guess-and-check mechanism

In the absence of an unobstructed view of the whole picture, the brain has developed the only valid mechanism of interpreting the picture, by guessing and checking. The missing details are "guessed" and the successfulness of the guess defines the outcome for the "guesser". As always in evolution, the wrong-guessers die before having a chance to produce offspring. In addition, as a group, the "right-guessers" (in other words, those who on average guess correctly) can push the "wrong-guessers" out of a favorable habitat, or suppress and absorb them as a group. Thus, this property gets consolidated and inherited. It develops and improves, gradually leading to abstract thinking.

We have learned from our own experience that our consciousness-vision is often mistaken. All sorts of optical illusions are based on this. This is actively used by professional illusionists, who deliberately make us guess incorrectly, either using purely optical techniques, causing "optical error", or by distracting our attention. Only a very narrow part of the picture focused by the eye at each moment of time, with everything else being "filled in" by consciousness. There are illusionist tricks that are based on this as well.

This mechanism of guess and check, which apparently exists in other higher animals, very probably was the first step in the (inevitable) emergence of abstract thinking and thus of the human mind.

By the way, humans are one the few creatures that can instantly increase their viewing range by standing up from a sitting position, and thereby see more details and check their initial guess. This may help to explain the evolution of a bipedal posture.

Perhaps this is exactly how the module of concepts emerged in its original, embryonic form, and was then strengthened and inherited over time.

Hypothesis 2. Revolutionary scenario: how fear and weapons turned ape into human.

It's tempting to suggest another, somewhat complementary scenario, drawing on some facts from anthropology. For example, the fact that the first humans made weapons and tools, which animals do not make. Therefore, it would be logical to assume that either weapons or tools were the first concepts invented by humans. In order to decide what was more likely, weapons or tools, let us recall some more facts. First, we know that one of our strongest emotions is fear, and emotions may very well play a role in the creation of concepts. The second is that creating a concept is not just difficult, but sometimes scary. Fear makes one think[7], look for a way out of a dangerous situation when something threatens your life.

And what can you think of under the influence of fear, especially in a situation of immediate and imminent danger? Weapons of course!

This is not just about the invention of weapons, but about the emergence of the **concept of weapons**, in other words the property that can be transferred to other objects. Chimps, according to various sources, can also use a stick as a weapon, and not just as a tool for knocking down bananas. And not just a stick. In captivity, they can also use stone throwing as a weapon. But unlike ancient humans, they do not make a spear from a stick, which they then carry with them at all times. Making and carrying weapons equalizes the chances for a chimpanzee in a confrontation with a lion.

For this to take place, something extraordinary had to happen. Something caused our ancestors to invent weapons and thereby undergo a phase transition in their brains, becoming the first humans. For example, this may have been the fear of imminent death in the paws and teeth of a strong predator, suddenly replaced by salvation as a result of a well-used stick with a sharp end that killed the attacker. It is quite possible that such an event forced our ancestor to realize (and remember) that it was the stick that saved him or her, and might do the same again. Therefore, the stick (already a spear) must be carefully selected, sharpened and carried at all times. That alone is a concept of weapons: a conscious selection and improvement of them.

---

[7] Of course, we also know situations where fear can, on the contrary, paralyze. But it depends on the degree of fear and the specific situation.

In addition, and this is no less significant, this would represent the emergence of the first logical operator in the mind— the operator of identifying a new concept – in this case weapons— as a means of defense (and, of course, attacks):

$$Stick = weapon \ (spear) \tag{2}$$

In the language of consciousness, this means the creation of a new (and in this case, the first) functional cluster responsible for an act of identification (operator of identification in terms of math), as well as a cluster of the first concept – in this case, weapons. Acquisition of this concept prompts humans to look for other things that can be turned into weapons.

In computer language, this is the creation of a new logical operator and, at the same time, a separate row in the database. Obviously, this action requires a substantial effort, is very energy-consuming, and is apparently accompanied by a huge resistance from the rest of consciousness, which now needs to give space to the newcomer in order to provide adequate resources for its newly acquired functions. It is certainly a phase transition.

In addition to the identification operator, there is a causal operator, which says that the death of the attacker was the result of a successfully used weapon. For a human this is obvious, but for an animal this is a highly non-trivial conclusion.

In other words, an apparatus of formal logic has emerged: an identification operator, a concept operator, and an operator (vector) of a cause-and-effect relation.

I dare to claim that the discovery (2) is more revolutionary and fundamental than $E = mc^2$ by Einstein, not only because of its immediate practical value, but primarily because it revolutionized our ancestors' brain and, therefore, made them modern, thinking humans. It was the beginning of the creation of the apparatus of formal logic. Firstly, because it created the first logical operator, and secondly, it prompted subsequent generations to look for new connections between seemingly unrelated entities and, in the process, create new concepts and models. Identifying an inconspicuous stick with the concept of a weapon is no less radical than associating energy with mass.

From the point of view of the survival of the species, this is followed by 1) the need to free hands, i.e. move on two legs, 2) the ability to move out of the jungle to the plain, where walking on two legs allows you to see further, and 3) the need to think about improving weapons and inventing new types of weapons, i.e. the need for more brains. Thus, we see that the invention of weapons not only made the ape a human, but also radically changed its appearance and status in the animal world. It made him a bipedal, large-headed, tailless, armed creature, the strongest in the animal kingdom, and lifted him to the top of the food chain.

As soon as it emerged, reason defeated physical strength and even experience!

Let me emphasize again – this is just one of the possible hypotheses. I was not there. I do not rule out that there was a combination of scenarios 1 and 2. Inheritance of the ability to create concepts is more consistent with scenario 1, as it is more extended in time than 2. It may have worked for several million years until something similar to scenario 2 happened.

The only thing that is clear to me is that the origin of reason is the same as the emergence of concepts. I believe that scenario 1 had to be a major contributor, whether in parallel with scenario 2 or without it. Type 2 scenarios can differ. The discovery of weapons could be just an accident. It is possible that similar "discoveries" began to be made by our ancestors on a regular basis long before the emergence of the *concept* of weapons.

There can be no doubt that the moment of the first invention and the beginning of regular manufacturing of tools and weapons by humans could be separated by hundreds of thousands of years. Besides, for this change to become irreversible, this new skill had to be consistently transferred, both to peers and to offspring. In the absence of a fully-fledged language, and because of the fact that people lived in relatively small groups, this was not easy.

The emergence of the apparatus of concepts served as an impetus for the emergence of language. At the very least, it meant that the need for language increased significantly. Each new concept requires notation. Assuming it is sound notation, the more concepts there are, the more complex sounds they will have to be labeled with. Over time, simple sounds will be insufficient. This is how language emerged. It allowed the effective transfer of knowledge and skills to offspring, allowed the union of large groups of humans, which in turn allowed defenses, hunting, and the raising of children to be organized collectively. It freed not only females and the elderly, but also tool makers. For a further brief history of our species, I can recommend Harari's book [12].

While we were still "reasonless", in common with all other animals, we were somewhere in the middle of the food chain, far behind lions, tigers – and even behind hyenas and jackals – because we had neither the strength nor speed to compete with them.

We jumped to the top almost overnight (on a historical scale) once we learned to think. This happened because thinking turned out to be much more economical and advantageous than simply chasing a deer or a mammoth.

Animals don't think. Upon receiving the signal, they instantly react. This reaction is encoded in the behavioral template created by evolution. How does a tiger, upon seeing a deer, decide to chase it? The tiger does not measure the deer's speed and trajectory, does not calculate how many calories will be spent on the chase and whether the chase is worth it, i.e. how many calories will be obtained as a result of catching the prey (which may need to be shared with offspring) as compared to the calories spent. The tiger does not have to solve the optimization problem. It doesn't even assess the risk of failure. Nor does it calculate how much energy it still has and how many failures it can afford before it will be unable to catch even a single deer. Nature, through natural selection, has solved the optimization problem for the tiger. Those who made the wrong decisions didn't survive. Only those whose decision templates were "correct" survived and passed on their genes. I use quotation marks here because "correct" is a very relative notion. What is correct in some circumstances may be fatally wrong in others, in another region or in another climate, for example. Entire species have gone extinct when the climate on Earth changed. And they died out not just because they were unable to withstand the heat or cold, a new level of the ocean or a drought. They died out because their decision-making templates ceased to correspond to the changed conditions and, for some reason, did not adjust quick enough to prevent their extinction. Those who managed to adapt survived. By contrast, humans were able to survive in almost any conditions, because they adapted their behavior almost instantly (on a historical scale). This was because they had learned to think.

Harari argues that the disproportionately large human brain and bipedal walking are mysterious and cannot be explained by evolution. However, they are perfectly explained by the revolution I just described. They both are the consequences of making, carrying, and using weapons.

In addition, once armed with weapons, it was no longer so critical to have a large, strong body, which also had to be fed. Therefore, the body size of humans decreased over time, as unnecessary, in contrast to the brain, the need for which only increased over time. As a result, armed apes (i.e. humans) have moved to the top of the food chain, and at the same time realized that their main weapon is their ability to think. Apparently this happened several million years ago. The invention of the bow and arrow, as well as the command of fire, completed humans' ascent to the royal throne of the animal world.

Thus, the beginning of the purposeful making of tools, which has lasted for many hundreds of thousands of years, should be considered the moment of the emergence of human intelligence. It's not just a matter of definition. Reason, in other words the ability to generate new concepts, or information, is what distinguishes a human from an animal. Purposeful production of tools is undoubtedly a sign of reason.

Language finalized this transition and turned humans into social creatures, i.e. modern Homo falsus.

From this moment on, the evolutionary development of our species, which works on a scale of millions of years, can be considered complete. A chain of revolutions has begun. The 2 million years that passed between the emergence of the first logical operator in the pre-human brain, which identified a stick with a spear, and the identification of energy with mass and the creation of an atomic bomb, is an instant on this scale. As we move forward, humans will modify themselves (and the rest of the animal kingdom) [16], unless we destroy ourselves in the process.

So, I hope we now understand how reason, which radically distinguishes humans from other animals, could have emerged.

Animals, too, have certain intelligence. This is evident from their behavior. They have behavioral patterns developed over millennia. However, it is also obvious that these models did not arise from intellectual generalization, but from accumulated experience and natural selection – individuals whose behavioral templates did not meet the challenges of survival in given circumstances died out.

Perhaps, in the absence of a fully-fledged language, cave paintings were one of the first mechanisms through which ancient people communicated information to each other and to future generations. Even if the paintings were simply reflections of the seen, creativity – the desire to capture what is seen – is in itself a sign of reason. This is in contrast to the plain curiosity observed in animals.

Here you probably want to interrupt me and point out the image of our ancestor, portrayed by anthropologists as a scary ape with a spear or club in his hand. Even the most terrible modern savage looks like a university professor by comparison. Am I serious in saying that our ancestors could reason? My answer is yes! First of all, two million years is quite a long time. And second, it would be good for us today not to forget who we are, despite our slightly more decent appearance.

Speaking of the age of modern humans, which is still an unsolved problem, let us do a little math exercise. Since each concept has a linguistic notation, the number of concepts that exist at a given level of civilization can be roughly estimated as the number of root systems and phrases in the language – which approximates the number of different words.

Knowing this number N at two or more points in time, t, for example, today ($t = T$) and Δt years ago, and assuming, for example, exponential growth $N(t) = a^t$, (where a is an arbitrary number) we can estimate how long ago modern humans evolved as the time elapsed from the first concept:

$$T = \Delta t \ \frac{\log N(T)}{\log N(T) - \log N(T - \Delta t)} \tag{3}$$

It has been claimed that in ancient Hebrew (about 3,500 years ago) there were about 23,000 different words, while in modern Hebrew there are about 33,000.

Substituting accordingly $N(T) = 33,000$, $N(T-\Delta t) = 23,000$, $\Delta t = 3,500$ into (3), we obtain $T \cong 100,000$, which, in order of magnitude, more or less agrees with anthropological estimates of the time of the emergence of language as 100,000-200,000 years ago.

Clearly, a more accurate count of concepts, a different growth pattern, as well as accounting for some historical events, can and should change this result. Thus, for example, if the number of words 3,500 years ago was only 25% higher (which implies slower growth), then instead of 100,000 years, we get 380,000. In addition, the number of words only remotely corresponds to the number of existing concepts. Therefore, of course, the above exercise should not be taken as a serious scientific calculation, but as a demonstration of the possibility of using elementary mathematics in this area.

Theoretical physicists can also contribute. A study conducted by Alex Gorsky and his co-authors indicates that the human brain is in a state consistent with a system in a near-critical state, close to a phase transition [17]. It indirectly supports my hypothesis that the emergence of any new concept (and this is what our consciousness is constantly engaged with, or at least will do) is a phase transition.

12. Miscellaneous issues

Here I will explore in a little more detail some of the aspects of consciousness discussed above. In addition, I will try to answer some questions of general interest.

Abstract thinking

When we talk about abstract thinking, correctly considering it to be a unique property of human consciousness that is absent in animals, we still make the mistake of assuming that there are even some people who do not possess it to a sufficient degree. We confuse abstract thinking with intelligence, which is indeed a property that varies between individuals.

In fact, abstract thinking is a fundamental property of human consciousness that follows from its very definition. In other words, human thinking and abstract thinking are one and the same.

As already discussed above, the concepts that are the constituents of human consciousness – and the reason for the existence of information – are themselves abstractions that often have little to do with reality, only existing in our imagination. Weapons and tools of labor, for example, do not exist in nature. They are concepts – abstractions – that we have invented.

All human consciousness is abstract in this sense. Of course, some concepts are related to reality through their values, but the absence of reality does not at all prevent a human from having a concept of this "reality".

Therefore, it is the ability to generalize and systematize information and generate theories and models – in other words, the ability to think abstractly – that distinguishes human consciousness from the consciousness of an animal.

In theoretical physics and other scientific fields, one of the key tools of analysis is a thought experiment. This is probably the highest manifestation of abstract thinking, beyond which one can hardly imagine anything else. The thought experiment is not only a replacement for the real experiment but is often an "experiment" that is fundamentally impossible to do in reality. Nevertheless, there are thought experiments the conclusions of which few people would doubt. For example, no one doubts that in the absence of external forces, a massive object will move uniformly and in a straight line, although it is impossible to verify this experimentally since it is impossible to remove other forces.

Thus, *Homo falsus* is none other than an ape who invents things that do not exist, or simply tells lies. He does it always, when he speaks, when he thinks, when he deceives, and when he "tells the truth". The latter can occur not only in the case of the delusion of an individual, but also for society at large. This is because society operates on fictions, which are by definition detached from reality. The individual fools himself by fantasizing, guessing, and imagining, thereby mistaking his fantasies for the truth. However paradoxical it may seem, sometimes these fantasies, especially in the exact sciences, lead to wonderful discoveries and allow us to achieve results that are quite in line with reality. This is like a stopped watch that, unlike a working watch, shows the exact time, but only twice a day. When one always lies, sometimes this lie turns out to be the truth, simply by chance, regardless of the desire of the liar.

Associative thinking

Associative thinking is a transfer of a property of one model to another, in essence completely unrelated to the first. The association is based on a shared concept or a common value among different concepts.

In his book *Search for Truth* [18] the renowned Soviet physicist A.B. Migdal described how an image of a circus performer riding a horse around in a circle led him to the solution of a problem in nuclear physics that he had been trying to solve for months. The problem was related to the orbital rotation of electrons in an atom (at least that was the model). One day Migdal attended the circus and saw a girl holding a bouquet of flowers while riding a horse in a circle. Her trick was to make an abrupt stop and release the bouquet, so that it flew into the audience by inertia. My memory of the story is that Migdal came up with the solution to the problem during his sleep, either the same day or the following day. In a dream, he saw the image of the girl and the bouquet flying by inertia. His brain pulled out the key idea from the image, which was the angular coordinate system! The transition to angular coordinates led to an instantaneous solution to the problem.

Of course, in order for this to happen, his brain had to constantly scan all the incoming information in search of an answer to the question. In this case, the association worked because of the common concept of circular motion in both pictures, which the scientist was thinking about at the same time – orbital rotation in an atom and a circus performer riding a horse in a circle.

Language as a weapon

As discussed here, the language that we humbly call a "means of communication", is primarily a means of deception. It deceives not only the people around us but also ourselves. The reason is that it consists of concepts, all of which are fictions invented by us, regardless of how well they are related to the reality they supposedly reflect. Language is just as powerful a weapon as the spear we once invented. It is a weapon for suppressing other people, subjugating them to the speaker's will and goals. As with spears, the first application of a human invention is often how to use it as a weapon, not just for personal protection, but also for the suppression of others. For as long as this trend exists, I don't think I can agree with the notion that humans have completed their transition from the animal world to another, more elevated realm.

One cannot deny the importance of language as a form of communication, a means of storing information, and as a tool for education. As a tool for creating and communicating concepts, language is fundamental to our civilization. It is also an internal stimulator of thinking. And yet it is a weapon which, like any other weapon, has its peaceful applications. The stick can be used as a tool and not just as a spear. I have no doubt that the discovery of the concept of weapons went hand-in-hand with the discovery of the concept of domestic tools. Atomic energy – initially intended only for making an atomic bomb (which, incidentally, has been a key component in preventing civilization's destruction) – was later used for electricity generation. But we must not forget that these peaceful applications were side effects of their creation, rather than the original motivation and the reason for their rapid development.

Any discovery can be used for different purposes, some for advancing civilized society and some for destroying it. They are always intertwined.

While it is language that allows the elites to deceive their peoples and keep them in subjection, it is also language that enables sellers to promote their goods and services, often misleading the buyer, despite all legal prohibitions. It is language that enables banks to provide valuable services to their customers and thus stimulate the economy. However, it also allows them to cheat and take advantage of their customers.

Language is the mechanism our consciousness uses to deceive itself and (as a result) forces itself to work on the invention of new (either false or true) theories, and on finding an explanation for everything in the universe.

The most dangerous application of language is its ability, under certain conditions, to brainwash entire populations. This allows authorities to instill into the mass consciousness concepts promoting war and the murder of entire peoples. The ability to influence mass consciousness is a much more dangerous weapon than any weapon of mass destruction. Just as the concepts of nation and state once allowed the elites to unite huge masses of people into nation states and unleash inter-ethnic wars, today's elites are successfully implementing globalist concepts with the goal of world dominance. The first attempts to unite groups of people larger than nations were world religions.

I want to draw your attention to the terrible danger posed by the monopolies of social media platforms, the news media and Hollywood. These monopolies are used to influence, control and manipulate mass consciousness. Illicit activities on Twitter, Facebook, YouTube, and other platforms that have influenced elections in the US show that these monopolies threaten the entire institution of democracy, and thus undermine the legitimacy of the elected governments. These activities have brought the country to the brink of civil war. They are controlled by individuals who are determined to destroy the US as a democratic republic and turn it into an autocracy. I was horrified by the world's reaction to the recent Covid-19 outbreak. Within an instant, all the achievements of the last 200 years' struggle for democracy, liberty, and freedom had been flushed down the toilet. Our so-called democratic governments seized the opportunity and became totalitarian. They were able to capitalize on the fear instilled in people by the mass media. As cover, they deployed pseudo-scientists who lied on demand of their masters. The newly born totalitarian governments, which only yesterday were calling themselves democratic, put their entire populations – hundreds of millions of people – under effective house arrest, mandated curfews and mask wearing, and broke businesses. By and large, the population complied. I am afraid that this was just a light test of how easily people can be subjugated and stripped of all their liberties under artificially created fear. I find it mind-boggling how fragile civilization is, and what may happen next.

Taking over Russian TV was the first and most crucial step for Vladimir Putin and his criminal gang in achieving the complete subjugation of Russians. As a result, this never-elected, quasi-educated, half-crazy, petty thief and KGB operative was able to quickly concentrate immense personal power in a country with nuclear weapons and unleash a bloody war against a neighboring country, thereby putting the whole world on the brink of complete annihilation.

These events clearly show us that the control of mass consciousness by a small group of people can lead to catastrophic consequences, not only for entire nations, but for the entirety of human civilization.

It is quite obvious to me that if humanity does not find a way to eliminate opportunities for this kind of mass brainwashing, sooner or later it will destroy itself. Thus, the language that created human civilization may become the cause of its extinction.

Thoughts-images. Wandering thoughts

Raw thought is a sensory image, a clip. Myriads of them constantly run through our sub-conscious. They are unconscious wandering thoughts. We are continually browsing or scrolling through them. This resembles a computer server working with a database in the form of images with indexes. The sub-conscious constantly snatches these images-sensations and either focuses on processing a particular one, if it corresponds to the current task, or drops it and moves on to the next. This browsing has no purpose. It is just how it is. One can say, of course, that the purpose is to keep refreshing the memory, since there is no other way to memorize. We do not have a stationary storage medium, so the only way to remember is to constantly browse. Or one could say that the purpose is to search for a particular thought. This would be the reverse of cause and effect.

This work of the sub-conscious, which never stops, even in sleep, is the animal component of our consciousness. Thus, our sub-conscious constantly scrolls through these thoughts-visions, as if wandering through them. Most of them remain unnoticed by us. Some are snatched by our consciousness and subjected to further processing. Sometimes we do this on purpose, and sometimes not. These thoughts are not just visual images or clips in isolation. Each is associated with an infinite number of indexes, markers corresponding to different emotions, themes, sensations, connections with other images, models, theories, and so on. One image can contain indexes of color, form, sound, smell, concepts, theories, emotions, and many others.

The uniquely human part of consciousness works on thoughts-images snatched from the sub-conscious. Its main goal is to identify concepts corresponding to the image. In the absence of an existing concept, it works on creating a new one. This work results in the verbalization of the thought-image. This identification and verbalization is what we call conceptualization.

As noted above, the visual component of information is its main part. This is why we are drawn to all kinds of beauty – scenic views, pretty people, beautiful buildings and paintings. This is why houses overlooking the water or the forest cost more, why rich savages build palaces, and why they like to visually demonstrate their wealth. That is why artists paint pictures, which we admire if they resonate with our perception of beauty.

A child eagerly absorbs everything around him through his eyes. He has not yet accumulated enough external information, nor has he sufficiently developed his internal concept generation and processing abilities. Over time, the proportion of external visual information decreases compared with its internal counterpart. For an adult, internal vision prevails over external. As a result, people who have lost their sight sometimes "perceive" more than those who are sighted. The higher the level of development of an individual, the smaller the proportion of external pictures in her perceptual and thinking processes.

This continual scrolling through thoughts also includes images of templates that are constantly spinning in our subconscious. We compare these with what we see. This is why, unlike a computer, we instantly recognize a familiar image, a friend, and also fall in love "at first sight" [15]. And this is why we sometimes, equally easily, make a recognition error. I am sure, like me, you have approached someone you thought you knew, only to realize it is a complete stranger you have never seen before. What has happened is that your consciousness made a recognition error by associating the person with the wrong template. It completed the "missing" details, and only when the actual details showed up realized that the differences with the template were irreconcilable.

Wandering thoughts are the thoughts that never get verbalized. They are just snatched from sub-consciousness without any purpose and then dropped. Normally, they don't have any impact.

How do you think? Language as the facilitator of thinking

The act of thinking is the intentional search for a specific thought and the verbalization of its outcome. It is not much different from a visual search for something. In order to conduct the search, you must formulate the question you are trying to answer, or the query sent to the sub-conscious. This is done according to the learned templates. The verbal formulation of the question or query forms corresponds to search filters. It is done according to existing templates, which makes the search purposeful, in contrast to the wandering thoughts described above.

As the renowned Soviet physicist L.D. Landau once said, a correctly formulated question is almost equivalent to a found answer. To find what you are looking for you need to visualize it first. After you have imagined a broad area where it could be found, you need to narrow this and conduct a search. Verbalizing the process facilitates the search and is equivalent to formulating a question, or a query.

I like to use shopping as an analogy. Your brain is like a huge supermarket. Wandering aimlessly along the aisles is the analog of wandering among thoughts. You browse through the items on sale and evaluate them one by one, asking yourself if you need the item that just popped up in your view. This is the opposite of a purposeful search. For this, you need to visualize what you are searching for (or verbalize it in the case of thinking), make a plan for the search, imagine what department your item could belong to, go to that department, and narrow the search to similar items. This is an analog of creating filters or queries. When you find the item (the answer), you need to grab it quickly (before it disappears), and add it to your shopping cart. This is an analog of verbalizing. The hardest type of search is scientific thinking. The main task is to create the correct filters, which is done according to the templates you have learned. In the scientific supermarket there are many departments and sub-departments and finding the correct one is a challenge in its own right. Creating logical constructs is analogous to narrowing the search.

Hence, language facilitates purposeful thinking. As a result, bilingual people are often asked what language they use for thinking. Many assume that thinking is the same as talking to yourself, so in everyday life, we use phrases like "think aloud", "let's think together", and "don't interrupt me while I'm thinking" (i.e. talking to myself).

I often ask friends, do you think while speaking? Many say no. Because what we consider thinking is purposeful thinking. The purposeless wandering among thoughts is not perceived as thinking per se. The reason we cannot purposely think while speaking is that such thinking requires verbalization. But we have only one speech organ that can be used. It's like having only one microphone at a debate. While one person is talking, everyone else has to wait.

Thus, language and speech are facilitators of thinking as much as they are tools of communication. The only difference is that communication is a conversation between two different people, whereas for thinking one has to talk to oneself. I see it as a conversation between different representatives of my "I".

Memory

To understand what memory is, recall a situation when you forgot a word, or someone's name, and tried to recall it. In an attempt to remember, you scroll through similar words. But when you think of a different word, even if it is very similar, you know for sure that this isn't what you were looking for. Only when the very word you were looking for comes to your mind do you know unmistakably that this is THE one. Sometimes, if you were unsuccessful in recalling, you leave attempts for a while, and then later the word suddenly comes up, unexpectedly, and, again, you know for sure that this is THE word. This means that the search continued in your sub-conscious, even though you decided to stop it. If you forgot the word, how in the world do you recognize it when it comes back, and how in the world, when a different one comes up, do you know that it is not THE word? The answer is that your body remembers the concept corresponding to the forgotten word. The concept is tabulated in sensations. In this way, your body remembers a whole range of sensations in which the forgotten word was tabulated. The word itself is just a notation of the concept. When you examine a different word, you don't experience exactly the same sensations that the forgotten word triggered. Even if it's very close, the feeling is not exactly the same. Only when the real forgotten word comes up does the whole orchestra of sensations and emotions play inside you, mixed with happiness from the fact that you found it. Thus, memory is a unique sensation corresponding to a given concept, word, or notion, encoded deep into the nervous system.

Memory is the association of sensations with a certain picture, thought, or concept. The more stable this connection, the more reliable the memorization. This explains how we recognize a forgotten word, image, or something else – we remember the range of sensations that it causes. In other words, our nervous system remembers the chemical portrait of a concept, of which the word denoting it is an integral part.

If, due to long non-use or for some other reason, the word corresponding to this concept does not come, the concept is not activated, and vice versa. It is activated only when the corresponding word (or phrase) is found.

Why don't modern apes turn into humans?

Firstly, who said that they do not? I say they do. This process can take millions of years, as we have discussed. The ability to "complete" the picture, to "think up" what you see from the fragments of the picture is obviously the germ of abstract thinking, of conceptualization. This ability may also be present in higher animals. Therefore, I can fully imagine that this process is going on as we speak.

Secondly, I tricked you with this question. What I really

meant to discuss was another topic: the emergence of human consciousness in our babies. We can observe this emergence almost as often as we want. Human babies are born without a single concept, just as animals are. The transition occurs between the ages of 3 and 5 years. Children who were adopted by animals when they were young, like the fictional boy Mowgli, have an opportunity to become fully developed humans only if they are returned to human society before a certain age, most likely by 5 years old. If this doesn't happen, then no amount of effort can make such a child into a fully fledged human being. He forever remains an animal.

It would be extremely interesting to observe this transformation in a human child in more detail. What happens to his brain at the time of the emergence of new concepts? Why can a human child, who at birth seems to be almost no different from an animal, be taught abstract thinking whereas animals cannot? What biological structure is responsible for this inheritance? Is it genetic or not? We differ from some primates by 1-2% of our genome. Therefore, the ability to learn concepts must be somewhere within this 1-2%. Can we see it by comparing our genes? Or if not genes, where does it reside in the brain and how is it passed on to children?

Why there are no bipedal apes?

This is not a frivolous question. Logically, there are two possible answers. First: all ancient apes that were bipeds turned into hominins. This would mean that bipedalism somehow inevitably led to the emergence of reason. One can imagine how this might happen, and I briefly touched on it above, but I doubt very much whether it actually occurred. Also, if this is true, why did all bipedal apes acquired reason almost simultaneously (on the evolutionary scale of time)? Why don't we see bipeds that are still in the process of becoming human-like? I believe this question almost eliminates the first answer as wrong. The second possible answer, as I already mentioned, is that reason is the cause of bipedalism and not its effect. I believe my reasoning that bipedalism was a consequence of the necessity to carry arms is quite plausible. One can object by pointing out that there is also a necessity to carry food and infants, which has no relation to reason. However, most animals accomplish these two tasks with the help of their teeth. So weapons are unique in this regard.

The only problem with the second scenario is the assertion by anthropologists that bipedalism emerged much earlier than reason did. Alternatively, anthropologists may be mistaken and must look for signs of reason that coincide with the beginning of bipedalism.

If we are to believe that bipedalism preceded reason, then the only thing I can imagine is that by standing up and thereby significantly increasing its viewing range, the ape experienced irreversible changes in consciousness, consisting of the ability to complete the missing details of the visible picture by guessing and then immediately (by standing up) checking the guess. One could object that such an ape could check its guesses by climbing a tree. But, firstly, there are few trees in the savanna where the bipeds went, and secondly, in the absence of adequate memory, the opportunity for instant verification may be critical. I personally don't like this explanation. While the guess-and-check mechanism is definitely there, I don't think bipedalism is the main contributor. It could play some role, but it is hard to imagine that it necessarily leads to the emergence of reason. In this case we would see fully bipedal modern apes among other apes. But we don't. I am rather inclined to think that anthropologists are mistaken and that upright walking is the result of the emergence of reason, and not the cause of it.

Another, intriguing claim that anthropologists make is that they can trace some types of early humans that didn't develop much and then regressed into apes. If this is true, and their heirs are still alive today and are not bipeds, this would completely rule out the idea that bipedalism preceded reason.

In any case, while the field of anthropology resolves these outstanding questions, I remain convinced that bipedalism is a result and not the cause of reason.

Sleep

Once we accept that consciousness is vision, the answer to the question of why we sleep follows automatically. We sleep because there is night on Earth and we are diurnal animals – our eyes see almost nothing at night. The night lighting that we only recently invented has not had enough time to affect our evolution. Therefore our consciousness at night has nothing to do but scroll through thoughts-images and recollect what it saw during the day. The little that we can see at night is not worth the energy spent on it, so it is better to just close our eyes. Consciousness almost immediately turns off. In some cases, we have a problem with it because consciousness keeps processing images that slip into the sub-conscious during the day.

Babies usually fall asleep during the day, since their consciousness-vision gets saturated and tired from seeing everything new. They absorb too much visual input.

When someone has a problem falling asleep, the best advice is "close your eyes". Consciousness will eventually follow by turning off.

What do we do in sleep? Why is our non-existent "daytime I" missing? It is because of what we discussed above – there is no input for fantasies and guessing and therefore there is "nobody" to associate this work with. But sometimes there is a certain nocturnal dreamer who creates very interesting night dreams for us. They are full of dramatic events featuring characters familiar to us from our daytime life and creatures completely unknown to us during the day. In a dream, all this seems perfectly logical, self-consistent and real. In the daytime, if you manage to remember the dream, which is rare, it all turns out to be complete nonsense. Now try to imagine what kind of nonsense our daytime fantasies look like to our nocturnal "I», which leads me to wonder, where is "truth", during the day or at night? Does it exist at all? I will discuss this question in chapter 13 about science and the limits of knowledge.

Many people tell me that they experience almost a "second life" in their dreams, where things happen that have nothing at all to do with their daily life. Moreover, these dreams can continue for months and years.

On the privacy of consciousness

The privacy of consciousness follows from the fundamental individuality of sensory experience and thus of concepts. Every concept, every word we use to denote it, is felt differently by each individual consciousness. What and how it feels is known only to the one who feels it, and no one else. We experience even the simplest concepts, such as color – mentioned at the very beginning – in different ways. I asked if we each see the same color in the same way. The answer is no, we don't. Here is a very simple proof.

I borrowed these samples from http://brainden.com/color-illusions.htm:

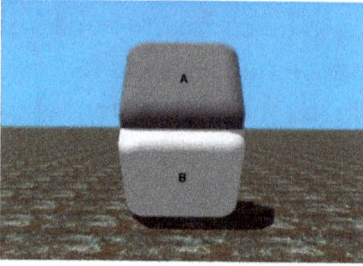

Sides A and B seem
a different color.

Horizontal strip between
A and B removed.

Fig. 1

Looking at the picture on the left it is difficult to believe that sides A and B are actually the same color. Side A is gray and side B is white, right? Wrong! The reason is that our consciousness-vision sees only the gradients, the differences in color, and not the "absolute" color, which does not exist at all. The brain uses the gradients at the transition between A and B to predict that B is in shade, and compensates accordingly. To verify that A and B are actually the same color one has only to remove the transitions, the gradients. On the right, I have removed the horizontal strips separating A and B without doing anything else. As you can see, they are indeed the same color, which is impossible to perceive in the original picture on the left.

One more example (same source):

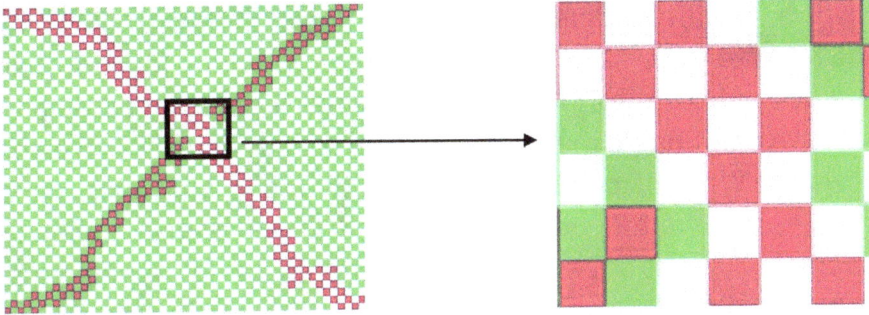

It is hard to believe that all the red squares in both strips are the same color

...Until you zoom in

Fig. 2

This is very similar. All the red squares, in both strips, are the same color. This can be verified either by magnifying the picture, as I did, or by cutting off two squares, one from each strip, and comparing them without any neighboring squares. They seem different because in one strip they are surrounded by white squares while in the other they are surrounded by green squares.

So we do not even need to compare our color perception with someone else's to confirm that the brain invents what we see. Our own brain sees the same color differently depending on the surrounding colors. The reason is that we only perceive a difference in color, a gradient, and not an absolute value, which in consciousness has no meaning at all.

As I mentioned above, even your doctor doesn't know if you are in pain, and may ask you to describe it or evaluate it on a scale of zero to ten. The same is true for more complex concepts, which have formed under the influence of our entire history, environment, language, profession, and so on. And even our own individual concepts change over time. Discussing this text with a colleague and a friend with whom I share the same background, education, and native language, I found that even for us some words have different meanings. For example, the word "illusion", which I often use here, has negative connotations for him that make it equivalent to "deception", whereas it does not cause any negative emotions in me. To me, illusion sounds more like an error of vision. What of people who speak different languages, have completely different backgrounds, and come from completely different cultures? Obviously the differences in perceptions, senses, and feelings could be so big that the whole meaning of what is said by one could be completely different from what is heard by the other.

It is no accident that in every language many words have a huge number of synonyms invented to reflect the subtlest nuances in meaning. But even these are perceived differently by everyone. Some words in one language cannot be adequately translated into another. This is because a relevant concept may not exist in another culture that corresponds to that language. A pleasant exception is the language of mathematics, but it covers only a small area of human activity. Thus, when talking to another person and trying to convey something to her, not only do we have no reason to assume that she perceives it in the same way as us, we can be sure that her perception will be different from ours. Sometimes it will be the exact opposite of our own.

The privacy of consciousness is fundamental. Fast forward to the future and imagine that we have the tools to draw a biochemical portrait of any concept for any individual. Even then, we will not be able to determine exactly what feelings a person has for a specific concept, least of all at any given moment in time.

Replacement of models

It has become a biological necessity for humans as a species to have a theory, a model, and an explanation for any occasion. That is why swindlers and professional liars – including a majority of politicians, priests, and journalists – have acquired such power over the masses. They satisfy the need of people to explain everything, and in the absence of reasonable explanations, they come up with whatever is most beneficial for themselves. Interestingly, people prefer the most implausible explanation to absence of any explanation. Similarly, they prefer a simple, albeit incorrect, explanation to a correct but difficult one. Thus the desire to have a theory for any occasion exceeds the desire to know the truth. Accordingly, it is understandable why simple but monstrous ideas like Nazism ("the Jews are to blame for everything"), Socialism (take and split everything), which contradict common sense, morality, and even a sense of self-preservation, so easily take possession of the masses and lead to heinous crimes.

A very unpleasant conclusion follows from this. Once idea A has been accepted by the individual, and especially by the masses, it is very difficult, and sometimes almost impossible, to replace it with idea B, especially if idea A is simpler. That is why new ideas are met with hostility, even in science – by people who, it would seem, by the nature of their activity, should be open to novel ideas, even if they are more complex. That is why, despite the rapid development of science in the last 200 years, the idea of God as the simplest universal "explanation" of everything is still alive and well. That is why it took the "Marshall Plan" to knock out the ideas of Nazism from the heads of an entire nation, or at least make people behave decently, and wait for the infected generation to die out. That is how the Putin gang managed to brainwash Russians whose brains suffered from the loss of imperial concept, with the collapse of the Soviet Union, and needed to be filled with something "new". The Putin gang cleverly used it by legitimizing the hatred toward the entire world, and especially towards the neighbors who don't want to live like Russians and/or with Russians.

This is also why we are witnessing a bitter war between right and left in America, from which there is no normal civilized way out.

It is clear that if idea A took hold of the masses earlier than idea B did, especially if it remained there for a long period of time, it would be very difficult to replace it with idea B. Indeed, in order to accept a new model, it is necessary to make a double effort. The first step is to destroy the concept that corresponds to an existing theory, and then acquire the new concept. As we have pointed out, even a single act of acquisition requires a lot of mental energy. Living with an old theory is easier.

When a young person begins thinking about politics, she has more or less equal choices between one direction and the other. You would think that this choice could be changed later if the person discovers new facts that contradict the theory she originally decided to accept. But no! Once the choice has been made the person begins to fight for it as fiercely as if she had been born with it. All new "facts" contradicting the accepted theory are ignored, or explained away. Every step will be taken to avoid changing the original choice (which initially may have been unintentional or due to a lack of education). And it requires incredible intellectual effort to change this choice later. It is especially difficult to replace it when the previously acquired theory is simpler than the one with which it needs to be replaced. The left-wing choice is always easier (it seems so simple – just take all the wealth and split it between everyone). That is why most young people lean to the left, and that is why the right is hopelessly losing this war.

Motivational multi-moves as a measure of cultural level of the individual and of society

The most direct way to measure a person's cultural level is to measure the number of motivational "moves" that they operate with. Animals operate with one move: caught/found (prey, food) → ate. The first humans had two moves: made a tool → caught prey → ate it. Their leaders had three: exert control over others → others make tools → caught prey → brought it to the leader to eat. For modern civilized people, a few extra moves have been added, such as getting an education (hence getting work, earning money, and only then eating), behaving decently (so that others trust you, and then the rest of the chain), behaving in a civilized manner (as expected in this particular society), and so on. So a person's cultural level is the number of moves that they operate with. Just as in chess, the more moves a player calculates ahead, the stronger they are.

For example, I did not immediately understand the multi-move plot of "climate change" promoted by globalists like Al Gore. Initially, it was unclear to me what benefits one could get from such a dubious narrative as "global warming". Not only because it is based on pseudoscience[8], and not because people should realize that being a power-hungry political animal, Gore is not a holy fighter for noble ideals (half of them know it anyway, and the other half have been brainwashed by the left-wing media and Hollywood), but because the very prospect of this narrative becoming a world religion seemed bleak to me. I was wrong. Besides, I did not immediately realize what benefit the elites represented by Gore could gain from this. It turns out that the whole idea is quite simple. Paired with power, it allows elites to steal huge national and multinational resources.

Under its cover, they can control the markets, tax the unwanted, and promote favorites. Instead of allowing honest competition, this will result in the elites deciding winners and losers, and almost literally deciding who is allowed to breathe and who is not. These represent many more moves than just plain stealing and eating. In addition, this is a play not just nationally but on a global scale. It's a goldmine! But to make a movie about it and even get a Nobel Prize for the movie are two additional moves. Outrageous, but brilliant!

Countries and societies can also be classified according to the number of moves they operate with. Russia and other totalitarian regimes, and medieval systems based on slavery, are three- to four-moves societies. The ruler forces his slaves to do everything, they bring him food, and he consumes it all. Industrial societies, "democracies", are five- to six-moves structures. The slaves are now "free people". No one forces them, but they voluntarily sell themselves for temporary use in the marketplace of work. The rules for sharing wealth at the top have also become more complicated. First, you need to win elections and gain power, and only then can you join the club of top shareholders. And to win elections you need to convince the slaves that you alone represent their interests. And before that you have to explain to them what their interests are and make them demand that their interests are taken into account. While countries such as Russia operate at the most primitive level (there are no elections, and the sharing of resources at the top is straightforward), in civilized societies there are at least two more moves (elections plus a more complex form of sharing).

By the way, the struggle between left and right in America (as well as everywhere else) is a struggle for a single additional link – freedom of the economy, freedom of the market. The left is for splitting the pie right away, while the right is for giving the economy the freedom to produce wealth first (and more of it), and only then split it.

Most communities are mixtures of cultural groups. Law and

morality tabulate the cultural level of the majority, of the dominant group. For example, in Russia, a minority of society (the so-called intelligentsia) has morality equal to civilized societies in Europe, i.e. five- and six-movers, while the state and the majority of the population are three- to four-movers. There is a struggle between them, which the minority loses. In the past, religion has played a huge role in establishing more multi-mover forms of morality (and law) than barbarian (two- or three-mover) communities, introducing prohibitions on killing, stealing, taking possession of another's woman, and so on. What is moral (legal) in less multi-mover societies becomes immoral (and illegal) in more advanced ones.

In other words, societies and individuals can be classified by their multi-mover level N, which is an indicator of both their level of thinking and cultural level. Animals correspond to one, the first humans were two, tribal humans are three, slavery-based and totalitarian societies are three to four, developing countries are four to five, industrial democracies are five to six. At the same time, it is clear that in every society there are people with a level of seven to eight or more. However, they don't dominate. Occasionally, because of their high intellectual levels, they occupy high positions within society, but their level has not yet reached dominance.

The idea of multi-moves came to my mind as a result of watching my relative, a very strong chess player, playing chess online[9]. He played blitz games and consistently won. Watching him, I caught myself thinking that in order to win one must calculate at least one move further ahead than the competitor. By that time, for several weeks, the question had been circulating in my head about what parameter can be used to classify human intelligence, the cultural level of an individual and of a whole community. And then I suddenly saw these chess games. The number of moves a person uses to create his logical constructs! After all, it's obvious.

---

[9] This is another example of associative thinking at work.

Savages think in two moves. Stole (took away) – ate. A civilized person thinks in multiple moves – got an education, a good job, did something valuable, earned money, and also invested in the financial market, which also needs to be studied first, and so on. And only then bought and ate. This is the behavior that a civilized society expects.

I remember the transformation of one representative of an uncivilized society into a civilized person, which took place literally in front of my eyes. It was a long time ago when I first met him. Let's call him R. This man had only recently immigrated to the States from one of the most uncivilized countries in the former Soviet Union. As a person, he was obviously talented, energetic, entrepreneurial, and eager to realize the American dream. He invited everyone he saw to become his partner in a joint business. The only feature he lacked was honesty. It was obvious to me. He gave the impression of a man who could not be trusted. He was still a member of that savage society that he had just left, one in which cheating on your partner is accepted as normal and not even condemned. Moreover, it is even praised and admired by many in that society. However, one of my friends didn't notice this feature of his character, and entered the proposed business partnership despite my warnings. Consequently, my friend lost a lot of money and was glad that not all of it was lost. Many years have passed and recently I was surprised to find that they are again in a joint business. To my question about how this partnership compares with his previous experiences with R, my friend replied that R is not the person he used to be. He has changed. He has become a completely different man, an honest and decent businessman who earns a lot of money, and even donates to charities. In other words, being by nature a smart person, R had realized that in this society, being honest and not cheating on partners and customers is much more profitable than cheating. He realized that trust in this society is a very valuable commodity that is transformed into business success, in contrast to the society of savages where he came from. Thus, he was able to increase his cultural level by a few points and become a respected member of society, and not a swindler as he had been in his previous life (in a country where being a swindler is a respected business).

One may object and say that people do not change. If he was a swindler there, he remained a swindler here. He just learned to hide it here, you could argue, while back there it was a matter of pride for him. There were decent people who were not swindlers even back there, among the savages. I agree. Despite all the recommendations, I am not a friend of R. But I definitely prefer a society where swindlers have to hide their deceit rather than brag about it.

Logical constructs: learning math and music as tools for advancing one's cultural level

Logical constructs are also multi-mover combinations, as in the previous example, and the most effective way to learn them is to study math. When solving a mathematical problem that is more complex than, say, arithmetic, one has to decide on a strategy to discover the solution. And the only correct strategy, as we were taught, is to divide the complex problem into a number of simple problems that can be easily solved separately. Hence, we learn how to make logical constructs. As with chess, and as with cultural multi-movers, the more steps one can master the better solver you will be. The construct is as much of a concept as any other concept. As we acquire the concept of logical constructs, we learn to solve not only mathematical problems but also any other logic-based problems. As a result, we advance our cultural level. Thus, it can be said that learning math, which is the same as learning logical constructs, advances human abilities. The more logic steps in the chain of a construct a person can master, the better solver he is, and therefore the more "human" he is. I'm not talking here about the qualities of a person. One can be a talented mathematician and a repulsive personality (and I know quite a few). I am referring to the level of abstract thinking.

Learning music is somewhat similar because it also requires long logical constructs, even though its language is different. Certainly, learning another language is beneficial on its own. As one of the famous said: "The more languages you speak the more of a human you are." But in addition to memorizing a large number of individual notes, music also requires the acquisition of its internal logic. In this respect I recall the famous story about young Mozart being accused of stealing a manuscript of Allegri's *Miserere* from St. Peter's Basilica in Rome. The music was so beautiful that copying it was prohibited, under threat of excommunication. This was meant to ensure that the music would only be performed at the Vatican, and only twice a year. Mozart, who was 14 years old, and his father came to Rome and attended two performances of the piece. After the first performance Amadeus wrote it down by heart, every single note of this 15-minute-long polyphony. At the second performance he secretly accessed the manuscript and made a correction. After being accused of stealing, he was able to prove his innocence. He was even rewarded with a knighthood by the Pope for making the correction that every expert agreed was necessary. Allegri had apparently made an error.

While Mozart was a natural genius, and not everyone can be like him, this proves that the human brain has tremendous potential, and that humanity still has a long way to go if it is to escape its own destruction.

13. Physics, the comprehensible world, and the effect of an observer

Exposing our delusions regarding consciousness, information, "I", and everything related to consciousness was not my original goal. I just wanted to understand how it works. But since deception, or self-deception, has become the main character of this text, I cannot leave science out. Moreover, thanks to this new understanding, some aspects of science and knowledge that I had not understood before became clearer to me.

Despite the fact that all the concepts that constitute our consciousness are abstractions that we invented, I hope you do not get the impression that they are all false or do not correspond to any reality. The vast majority of the simple concepts, such as apple, parents, and tree given earlier as examples, are quite consistent with reality. It is impossible to argue with the fact that each of us was born by a mother and father, that an apple, although it is just a collection of molecules, really exists, that specific trees also exist, despite the classifications we have invented and despite the non-existent properties attributed to them. Yes, in the real world there is no color, taste, smell, or Beethoven's symphonies. But they are in our minds, and this is also a reality. Yes, we came up with money, nations, states, corporations, god, beauty, virtue, evil and much more, which did not exist in nature before us. But all this now exists, even if only in our minds, and has a decisive influence on our lives.

We came up with physics, mathematics, chemistry, and medicine in a similar way. As a result, airplanes fly and transport us to any part of the globe in a matter of hours. Cars, trains, electricity, the internet, and mobile communications operate. Nuclear energy also works, and the atomic bomb, unfortunately, explodes. And although medicine is still infinitely far from being called a science, doctors treat us and prolong our lives. Therefore, we cannot deny all of this reality we have created. But without denying it, it is essential to remember which concepts correspond to which realities, which are useful abstractions, and which are false or even harmful delusions.

Any of our concepts – despite their abstract nature – may be useful or harmful. As a rule, harmful or useless concepts sooner or later disappear from use. Religion, or at least its most odious components, is gradually dying out. Scientists no longer count the number of angels that fit on the end of a pin.

The once accepted concepts of racial and gender superiority are dying out because they're wrong and harmful. All this happens as part of the wider political struggle and, as a result, the "fighters" for the equality of blacks have turned (not all of course) into fighters for discrimination against whites, and the "fighters" for women's equality into fighters for discrimination against men. And now, instead of establishing equality, they are fighting for a new inequality – in the opposite direction. And everything is upside down again, but in a different way.

I am quite sure that if humanity is to escape self-annihilation, nation states and national borders must ultimately be abolished. I understand some of you may be outraged by this idea. However, let me remind you that only a few thousand years ago there were no nations and the Earth belonged to all humans and animals equally. If we are to end wars and killing each other over a piece of land or any other material "valuables" we will have to learn to share. When the first colonists came to America they did not have to obtain visas or green cards. And not just the first colonists: 150 years ago everyone could come to America freely. It belonged to all humanity. Why do we now close our borders to others? If we do not want to share what we have earned, that is our right, in my opinion. Let us not provide automatic, free "benefits" to newcomers, just as nothing was provided for free to the first colonists and to many generations that followed them.

As a matter of fact, what faced these early settlers was natural selection. Knowing that there was nobody there waiting for them with a piece of pie, the potential newcomer could only count on his or her own survival skills. Thus, only strong, adventurous, self-confident, entrepreneurial people dared to make this journey. This early natural selection explains American success as a society. Why don't we keep this by eliminating all social benefits for newcomers but allowing free immigration? Don't get me wrong, I am not so naive as to think that this is all there is to immigration reform, which incidentally everyone agrees is necessary. This is simply an argument that needs to be taken into account while contemplating such reform.

But let's leave dirty politics aside and move on to "pure" science. Let's start with physics. What concepts of physics correspond to reality? We start with the most fundamental ones.

Time

Time exists only if there are events, i.e. changes in some parameters. If there are no events, then there is no time. I think this is obvious. If it is not obvious to anyone reading this, let's prove it by contradiction. For simplicity, imagine a world in which there is only one changing parameter. We call this change an event. How much time passed between two events in this world? Remember that there were no other events between these two events. Therefore, the answer is none, time did not exist. Since nothing has changed there is no way to measure this "time". It just wasn't there. In order to measure, we would have to have at least an imaginary clock, and its imaginary hand would have to turn. But it couldn't turn because this would be an event, and we agreed that there were no other events. Consequently, time as an absolute quantity does not and cannot exist.

Instead, there is a frequency of events. But even frequency does not exist if only one quantity changes. We cannot say how often it changes if there is nothing to compare with. Therefore, frequency can only be relative. Let's take the frequency of the most frequent event A as the unit of measurement, i.e. it is equal 1. Then the frequency of any other event, say B, which is t times slower than A, is $1/t$. Here t is exactly the time that has passed between the changes in B (during this time, event A has occurred t times).

This agrees with Einstein's theory of special relativity, which says that there is no time in the photon's reference system. This is because if you fly with a photon, which is a carrier of interaction, nothing can change for you. Therefore, there is no time there.

Moreover, time cannot be continuous. It must be a discrete quantity. To have continuous time one must have events of infinitely small time between them, or of infinite frequency (and, by the way, of infinite energy as we will see below). This makes them impossible since they would lose the whole notion of an *event* if this were the case. An event by definition is a discrete entity. Therefore, time must be a discrete entity too. In other words, for the events or the changes to exist there should be a limit to how fast things can change in our eventful world. Please note that I am not talking about Planck time, which comes from quantum mechanics. The conclusion is just a logical one, without any reference to quantum theory. But it is good that they are aligned.

Thus time is neither fundamental nor continuous.

Space

For the same reason that there is no time without events, there is no *observable* space without motion. It should be emphasized here that since physics is a mathematical description of *observed* phenomena, the presence of at least a hypothetical observer is a necessary component. What cannot be observed, even in principle, has nothing to do with physics as a science, and therefore is not the subject of our discussion.

To observe that two points are separated by space it must be possible for them to exchange a signal. Moreover, the signal must propagate at a finite speed. If the speed is infinite, this is equivalent to zero distance and there is still no spatial separation.

For the same reason as for time, there is no continuous physical space. Continuous would mean that two points of space can be infinitely close to each other. This is impossible since it would violate either the limitation on the minimal time interval or the limitation on maximal speed. Again, this conclusion comes neither from quantum mechanics nor from the existence of the Planck scale. Vice versa, the latter could be a direct consequence of this conclusion.

Thus, neither time nor space are fundamental properties of our world, but are just fictions that we have introduced for convenience of description. They are derivatives of more fundamental quantities, such as motion, events, and the finite speed of a signal.

Dimension of space

One might think that at least the dimension of space is a fundamental quantity, though this assumption seems deeply suspicious in view of what has just been said about space itself. Indeed, not only is the dimension not fundamental, but we can change it as we please. The culprit is again our own consciousness. For motion in D-dimensional physical space to be observed, the observer must be at least (D+1)-dimensional. To understand this statement let us conduct a thought experiment and initially assume that we are one-dimensional and live in one-dimensional space, i.e. on a line. All events and motion also happen on the same line. Signals from them propagate along the same line. Obviously, all that we will see is zero-dimensional space, i.e. one point, and no motion. We will not be able to see or measure any distances. Only a 2D or higher-dimensional observer can observe the motion in 1D space. Similarly, if we now move to 2D we realize that if we are 2D ourselves, the only motion we can observe is motion along a line, i.e. a 1D motion. It would be impossible to observe actual 2D motion for the same reason that we could see only a 1D projection. Only a 3D or higher-dimensional observer can see a 2D motion. From here it is easy to guess that the D-dimensional world can only be seen by a (D+1)-dimensional observer. How do we see three-dimensional motion if we ourselves are three-dimensional and all we can see is its 2D projection? The answer is with difficulty and by guessing.

That is why we have our consciousness-vision, which in this sense is our fourth dimension. It only guesses what is there, behind the two-dimensional projection. Sometimes it guesses correctly, sometimes not. Unfortunately, it cannot see the three-dimensional world in its entirety, as we see the two-dimensional world. But as in the case of a 2D observer trying to observe a 2D world, we have to crawl through our 3D world and guess its spatial properties. Slightly spaced eyes, through which signals of different phases enter our consciousness, are of some help, but our ingenious consciousness is the main tool. It draws 3D from the 2D image it receives.

For example:

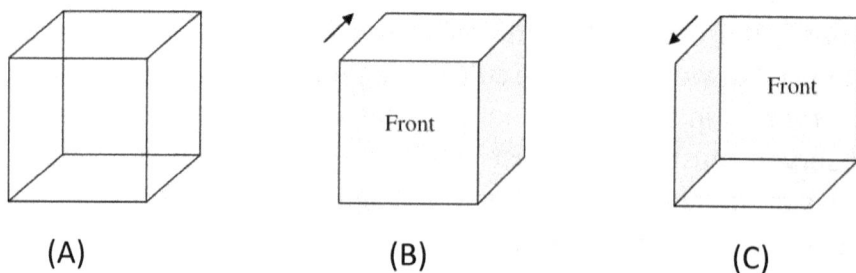

(A)   (B)   (C)

Fig. 3

Looking at the two-dimensional picture (A) one cannot decide which one of the two possible cubes, (B) or (C), it corresponds to. Notice that to make it easier to interpret I did not add anything, except for the arrows, labels and shadows for the especially bad guessers, but rather removed some lines. This is exactly what our consciousness-vision does. Look at image (A) again and try to see (B) and (C) in it at the same time. For some reason that I don't quite understand, I see picture (C) in two different ways. One is as I initially marked with the word "Front" implying that the front side is closer to me, and the other one is with the word "Front" being written on the internal back side of the corresponding box. My guess is that it has something to do with right- and left-handedness, or with cooperation between the right and left hemispheres of the brain. Perhaps some other people (perhaps left-handed people) would have a similar problem with (B).

We don't see 3D images. All of our image media – paintings, photographs, movies – are two-dimensional. The exceptions are sculpture and theater, but here we first convert the image into two dimensions, and then habitually "guess" its three-dimensional meaning. Walking around the sculpture, we only make sure that our guess is correct, but we still see only in 2D. To see in 3D we would have to see the entire sculpture simultaneously from all sides, just as we see the entire 2D picture at once ("from above").

Quantum phenomena and probabilities as effects of an observer

When physicists talk to a layperson about quantum phenomena, they take a long time to prepare the listener for the fact that there is nothing to be terrified about, that quantum phenomena happen only in the micro world – which the listener will never see anyway – and that in our real, macro world everything is good and "classical", so there is nothing to worry about. They imply that the listener is about to learn top-secret information, that they are about to be admitted to the "holy of holies". However, the layperson is still unlikely to comprehend anything, since this knowledge is only accessible to those who are particularly privy, the PhDs among PhDs, and not even all of them. Finally, they are told that "observables" in this mysterious world are of a probabilistic nature, that there is nothing certain there, and that nature at this level operates not with real objects but with their probabilities. Though again, in their classical, macro world everything is strictly defined and determined. So go in peace, say the physicists, and don't ask any more questions about quantum physics. You've been told too much already.

For all the non-physicist readers, I have a surprise for you: you were conned. Furthermore, several generations of physicists have lived with this lie and passed it on to others. In fact, everything is much simpler. There is nothing sacred about it, and the micro world has absolutely nothing to do with it.

Just look again at the pictures in Fig. 3. (A) is the optical image supplied to our eyes, whereas (B) or (C) is what we *see* as a result of the work of consciousness-vision. Nevertheless, who decides which of these two we end up seeing, (B) or (C), and how? Both are equal *possibilities* for how the two-dimensional projection (A) could have arisen. These are the very "quantum" probabilities you were told about. In this case, I drew two. In general, any two-dimensional picture (which is what is supplied to our eyes) is a projection of an infinite number of possible three-dimensional "realities" (quantum states).

It is our consciousness that removes the ambiguity, every second deciding in favor of one or another possibility. Which one it prefers depends on our experience. Naturally, we do not notice this process, since the removal of uncertainty occurs instantly, in the sub-conscious, with the exception of cases when the guess turns out to be incorrect. In this case, consciousness quickly corrects the picture and everything falls into place. Although sometimes we still notice the "glitch". Thus, the whole "mystery" of quantum physics lies in the uncertainty of which one of the many possible three-dimensional realities resulted in the two-dimensional projection we observe.

Speaking of "glitches", I cannot help but recall the well-known *déjà* vu phenomenon when you have a feeling that you already experienced what you just saw, only in the past, while you know for sure this could not be the case. In fact, this is just a direct result of what I just described. Only in this case the "glitch" was significantly delayed. Your consciousness has struggled for too long (perhaps a fraction of a second) to make sense of the scene, and as a result you saw the scene twice. So if you ask me when you saw this same scene before, my answer is – a split second ago.

Physicists routinely refer to the concept of "observables" and a hypothetical "observer". However, the latter is reduced to a kind of convenient, handy servant. When needed, he can be taken out of the physicist's pocket or, when he is not needed anymore, put back and forgotten, as if he had never existed at all. His "point of view" is never really taken into account when describing physical phenomena. Perhaps the time has come to rethink this vicious cycle and include the observer in the picture of the observed world, at least in order to understand why it is observed in this and not in another way, and stop ignoring the word "observed". It must become an integral part of the equations by which we try to describe what we observe.

I think it is clear from the above example that there are no "probabilities" in nature. Probabilities are only an appearance. They look like that because the reality behind them is hidden from our view. They exist only in our consciousness. It is consciousness that chooses which of the possibilities (in the example above, (B) or (C)) should be accepted as reality. It is not the uncertainty of reality that determines how closely the choice corresponds with reality, but the uncertainty of the choice. Until we know reality, both choices, (B) and (C), seem equally probable. But once further verification is performed the ambiguity is removed. Before verification, consciousness is forced to assume one of the possibilities, and in this sense it is a probability. In other words, probabilities are a sign that we are not seeing the whole picture, so we must assign probabilities to different interpretations.

The so-called "principal" uncertainty of reality is nothing but the uncertainty of observation. As soon as a real observer is included, uncertainty is removed. And we know this quite well from quantum-mechanical experiments. So far, we've just misinterpreted them.

In general, when you are told to replace deterministic views with probabilistic ones, you can be sure that you are being deceived. Creationists often suggest that the evolutionary emergence of life, and in particular intelligent life, was highly unlikely given the astronomical number of factors that would be "necessary", and it must therefore be an act of God. I call this trick the lottery trick. Its essence is to change the reference system. Imagine a lottery, say Power Ball. By definition, the probability that someone will win a jackpot is 1, or 100% – the lottery is held until someone wins. Now look at the result from the point of view of a particular player, John Smith, who is the winner. What was the probability of winning from his point of view? It was almost zero. From his point of view, winning seems tantamount to an act of divine intervention. The reason for the discrepancy is the different reference systems. In the first case, the probability is calculated relative to all participants, and in the second, relative to a particular one. You can apply this trick to yourself, and calculate the probability of you being born, exactly from your parents, exactly in that particular place where you were born, and exactly on that particular day. Don't forget to take into account the chances of your grandparents and great-grandparents and all your ancestors meeting each other, including the fact that a particular remote ancestor of yours, who lived thousands years ago, at a particular moment decided to move north and not west. When enough factors are involved, you will get a probability much smaller than the probability of the emergence of life on Earth mentioned above. If you add enough factors, you can arrive at any arbitrarily small number that may be even smaller than the probability of the universe's birth. As you can see, your chance of being born the person you are is zero. You either cannot exist, or it's an act of God. (I hope the reader understands my sarcasm here.)

Given the level of biological complexity that the brain and nervous system of animals has reached, and in the light of what we learned in the chapter on the emergence of human consciousness, the probability of the emergence of humans is equal to one[10]. This is indirectly confirmed by anthropologists' assertion that at one time more than 20 different types of humans existed on Earth alongside our ancestors. Evolution has left us as the only survivors, while the rest have gone extinct. This is a historical question rather than a biological one. Given the latest events in our own history, we can say that our survival as a species is also far from being determined.

Thus, the illusion of the probabilistic nature of reality arises from the fact that the 3D observer can only observe the 2D projection of reality. Upon observing the projection, the observer perceives all possible "realities" that might have resulted in the projection as alternative possibilities. Thus, we are actually talking about consciousness and ambiguities in the interpretation of events that it observes.

This is the **effect of an observer.**

---

[10] While, according to probabilistic considerations, these same humans could not possibly have emerged.

Generally, if the result of an experiment depends on whether the observer is present, there can be no doubt that this result is an illusion of our consciousness. It is not objective reality. Analogously, the presence of probabilities in physics formulas can mean only one thing – that they describe not a real phenomenon, but rather a perception of it.

Particle-wave duality

Particle-wave duality is another effect of an observer. It is a result of observing the same phenomena in two different dimensions. Let us consider a particle moving at a constant speed in a circle (see Figure 4(a), below). This is a 1D motion. It can also be viewed in projection 4(b). In this projection, instead of a particle rotating in a circle we see a particle oscillating between two points that correspond to the north and south poles of the circle. Now let us add one more dimension, for example time, as shown in Figure 4(c), and look at this motion in 2D. As we see the 1D oscillating particle becomes a wave in 2D.

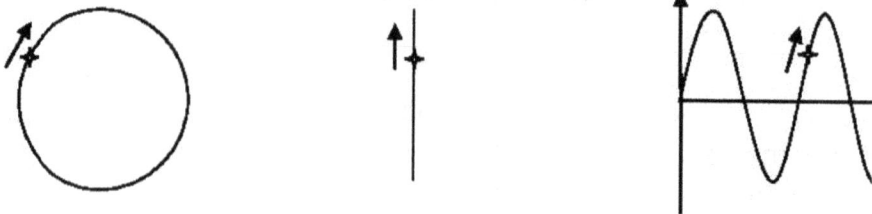

Fig. 4

Thus, the same phenomenon can be viewed as a particle or as a wave depending on the dimension of space it is viewed from. It seems that the number of dimensions is a property that our consciousness freely manipulates. Duality is the result of looking at the same phenomena from (literally) different perspectives. As explained above, due to the spacing of our eyes and because of the phase difference between optical signals coming to them, we see a quasi-3D picture. Closing one eye makes the picture purely 2D. Thus, by observing, we change the dimension of space, which literally depends on our point of view.

But let's return to physics.

Mass, energy, electric charge, elementary particles, and quantum fields

Energy and mass are as much derivatives of motion as are time and space. It is understood that energy is a mysterious source (cause) of motion and change, and mass is the energy of compensatory motion[11], or a measure of inertia. Thus, there are two opposite energies – the energy of motion and the energy of "inertia" opposing it, i.e. resistance to motion, which always appear in a pair, so that the resulting energy is equal to zero, without disturbing the rest of the stationary world[12].

It is quite obvious to me that the most fundamental definition of energy is its definition in terms of frequency, by analogy with Planck's formula: $E=hf$ where f should be understood as the frequency of a wave or oscillation of the corresponding reality.

But, in fact, energy is a fiction. Accordingly, derivatives of energy, such as force and momentum, are also fictions.

Ask a physicist what an electron is, what electric charge is, and why all electrons are identical. An honest scientist will admit that we have no answer. Otherwise, you will hear a lecture about interaction of charges, about electromagnetic forces, and so on.

---

[11] Thanks to Einstein, we now at least understand that mass and energy are the same: $E = mc^2$ (up to a factor equal to the square of the speed of light).
[12] This was correctly noticed by Newton and is reflected in his third law of motion.

The truth is that all we know are the relations between quantities. We know how electric charges interact with each other, but we cannot say what they are and how they came into existence. The same is true for the spin of a particle, for the color of quarks, and all other physical quantities. No attempt is even made to explain the quantum fields that supposedly produce all these quantities.

Unfortunately, all these quantities are fictions, concepts, about which we can say a lot except for what reality they correspond to, where they came from, and what they are. The most we can say about them is how they relate to each other. In fact, we have achieved tremendous success in this regard. The laws and equations of Newton, Coulomb, Maxwell, and Einstein describe the relationships between these fictions with amazing accuracy. When used in engineering they produce tremendous results like planes, ships, electricity, and the atomic bomb. In other words, **physics is the science of the relations between fictions**.

Interestingly, all of these fictions obey various "conservation laws" (conservation of energy, mass, momentum, charge, and so on). This only confirms that they are fictitious quantities that belong in the physicist's "kitchen" rather than in the real world, if such exists. Just like coordinate systems, just like absolute space and time, they are handy utensils for physicists to do their work but should not appear in the final dish. That's why we have these conservation laws. Simply put, these laws say you can invent any fiction and deal with it, but if you want your description to match reality, the fictions should eventually annihilate each other (or be equal to zero if you sum them up).

Despite the discoveries of Newton and all the successes in the application of his formulas, the origin of gravity remained a mystery for a long time. The puzzle was solved thanks to Einstein's discoveries: his equations of general relativity, their solutions, and, ultimately, experimental confirmation, primarily the redshift[13]. However, if we were four-dimensional, we would not need the Hubble experiment to understand that both the Universe and space are expanding. We would see it with our own eyes. We do not see it because we ourselves are three-dimensional. But we feel it through the gravitational "force". As Einstein guessed, gravity is not a force at all, but the result of inertial resistance to acceleration. What we perceive as gravity is actually the same force that pushes us backwards into the driver's seat when we step on the gas and accelerate. But unlike the car example (Einstein imagined an accelerating or falling elevator), we do not see a three-dimensional expansion of space, since we ourselves are in it and we consist of it.

Thus, gravity is another result of motion. I have no doubt that all the other so-called "forces of nature" are inertial forces, i.e. the result of compensatory resistance to movement. It is only necessary to understand in what space and what dimension this movement takes place.

Until we understand the reality to which physical concepts such as energy, mass, and charge correspond, we cannot explain or calculate, based on fundamental principles, the mass or charge of an electron, the speed of light, and the fine structure constant. We don't even know how to approach these issues, although it would seem that this should be one of the main goals of science.

From the reasoning I have just given, and from many previous ones, it follows that the only physical quantity that can really be considered fundamental, and that can be relied upon in obtaining all other properties and quantities, is motion. Motion

---

[13] For non-physicist: redshift is a Doppler shift of light's frequency towards lower (redder) end, which means that the sources of light (stars) are moving away from us. If the shift was blue, i.e. towards higher frequencies, it would mean that the sources are moving towards us. In his experiment Hubble confirmed that the shift is red and thus the Universe is expanding as predicted by Einstein's general relativity.

is the reason why we have space, time, energy (and mass), charge, and all the other "quantum numbers", as well as "forces of nature" and "conservation laws". So, I believe that the main task of physics in the near future will be to calculate charges and constants based on first principles, i.e. on motion. And, of course, we need to understand where motion itself comes from and how fundamental it is.

On the "miracle" of mathematics

I cannot help but dwell on the myth that mathematics somehow − miraculously, inexplicably − corresponds to the observed world, and that this almost proves the existence of the Creator. I speak as an unbeliever who is offended on behalf of the Creator and who wants to defend Him from diminishment. While the usefulness of math is indisputable, the "miraculous" part is a myth, because mathematics is simply a language of quantities, which we were very lucky to invent, and which makes it easier for us to calculate large quantities, including infinities. The whole "miracle" of mathematics is that one apple plus one apple is called two apples. Just "called", and nothing more. It is not even "equals to" which could be considered a miracle if, for example, the number 2 had any other, independent definition. But it doesn't. It's just a naming convention: let's call 1+1 number 2, 1+1+1 number 3 (and therefore 2+1=3, but this is not a miracle at all), and so on, to infinity. Then we call the inverse operation subtraction; repeated addition, multiplication; its inverse, division; repeated multiplication, a degree; the inverse of it, the root; etc. All of this is then generalized to integrals, derivatives, in all kinds of dimensions, sets, manifolds, and so on. There are no miracles. Everything else comes from this. Mathematics is just a language, a naming convention. It is the language of computation, logic, and certainly of physics. Like all other languages, it is made up of concepts such as mathematical operators, geometric and logical constructions.

110

In fact, any mathematical statement is a truism (if it is correct, i.e. if all transformations were made without errors) that is no more miraculous than 1+1=2. Yes, sometimes it can be very difficult to do. Proving Fermat's theorem is a remarkable recent example. Nevertheless, this does not change the essence of the matter. Mathematical statements are either wrong or truisms, no matter how difficult the proof is.

Laws of physics

A myth that's worth considering is the so-called simplicity, elegance, or beauty of the laws of physics, from which some conclude that they must be "intelligently designed". At the same time, the laws of physics are called "Laws of Nature" and are considered as something external to the material world itself, as if they exist independently of it, in some other world [19]. For an attentive reader of this text, it is not difficult to answer the question of where these very "laws" exist. The answer is that they are in the same place as any other piece of information – in our consciousness, and nowhere else. In addition, it goes without saying that the very concepts of "simplicity" and "beauty" are so subjective that the whole claim is nonsensical. What is simple for one person may be extremely complicated for another. What is beautiful for one person might be ugly for another. Personally, I find *some* laws of physics elegant. Definitely not all of them. I prefer to formulate the question a bit differently – why are *some* of the *laws of physics* simple and elegant, and how did it happen that some of them adequately describe the observed reality (otherwise planes would not fly, ships would not sail, and our smartphones would not work)? By adequate, I mean that they make correct predictions that can be verified in an experiment.

The laws of physics are, in fact, a reflection of the fictitiousness of the above-mentioned "physical quantities", which were introduced by us for the sake of convenience to describe

111

observed phenomena. It is evident that if we want to maintain the connection with reality, the latter cannot rely on the intermediate fictions we have created. Just as coordinate systems or gauge transformations are the "internal kitchen" of theoretical physics and cannot affect observed quantities, fictitious entities such as energy, momentum, mass, and charge must disappear from the big picture, even if they help in some intermediate calculations. This is where the "conservation laws" come from. "Conservation" is the same as nullification. If a quantity, say energy, is conserved, this means that the input amount of this quantity equals the output. Thus, the difference between input and output is zero. This is very well known. Therefore, I call all "conservation laws" by one common name – the **law of nullification of fictions.**

The eventless world is easily described by a single formula for its partition function:

$$Z = e^S = 1 \qquad \text{(and thus, action } S = 0) \qquad (4)$$

The world of events and motion in which we live is a little more complicated:

$$Z = \int_{\aleph} dU \, e^{\sum_i S_i (U)} = 1 \qquad (5)$$

Here, the integral is taken over all physical concepts U (including quantum fields) that belong to the space of all concepts, $\aleph$, which in turn can be either an adequate or inadequate product of our consciousness. The sum in the exponent goes over all values i of a concept U where $S_i (U)$ is the action corresponding to a particular state (value) of the concept. As we have already seen, even if something is an inadequate product of consciousness, its presence in (5) should not lead to discrepancies with reality, because by this relation the correspondence to reality is guaranteed by the effective absence of fiction in the end result.

Strictly speaking, (5) is the same universal law of the conservation of fictions that "governs" our world of motion, and through its end result of 1 connects us with the motionless world (4), in which it is embedded. Whether we see manifestations of the motionless world in the form of black holes, where any motion stops, including time and space, or whether we ourselves live in one such black hole, we do not know.

All conservation laws, all symmetries, and all "forces of nature" are hidden here. The apparent simplicity of (5) ends when one specifies a concrete subset of ℵ and tries solving equation (5) for the specific subset.

When someone tells me about the simplicity and mathematical elegance of laws of physics, I want to ask, have you tried solving three-dimensional or four-dimensional QCD (quantum chromodynamics)? Apparently you have not. Otherwise you wouldn't make such a claim. I did. For simplicity, I have found an exact solution to the 2D QCD partition function on arbitrary compact surfaces. It has resulted in a mathematically compact formula [20][14]:

$$Z = \sum_r d_r^{\,2-2g}\, e^{-\lambda Cr\, A} \tag{6}$$

But its simplicity is a visual illusion. Let me note that the sum is taken over all irreducible representations r of the gauge group $U(N)$, or $SU(N)$, and formally it is not much better than the original integral. The remarkable thing is that it was possible to separate geometric properties, such as area A of the surface and its genus g from purely algebraic properties of the group (here d_r is the dimension of the representation, C_r is the second Casimir

---

[14] I cannot help but notice that the compactness of this formula is so fascinating that E. Witten published it in [21] without any reference to my paper [20], as if he had found it independently, and had not learned it from my preprint I sent him one year before. I like to believe that he just forgot about it.

operator in this representation, and $1/\lambda$ is the strong interaction constant), but it is not so easy to understand the true meaning of this formula, and only thanks to the efforts of numerous authors (more than 300 publications to date), the first of whom was David Gross [22], the real meaning of this formula became clearer. And this is just two-dimensions, which has a very weak relation to reality. The solution in 3D and 4D, despite my optimism [23], has not been found so far.

Yes, it is true that some phenomena allow for a very compact description. And the question of the reasons for this is not as silly as some may think. The conclusion about "intelligent design" does not satisfy my curiosity, and contradicts my personal idea of beauty and simplicity. The rational answer to this question is the real task of science, and should open a way to understanding the world in which we live and our real place in it as carriers of reason, capable of knowing.

The question of whether the world is comprehensible is essentially a question about the existence of reality outside our consciousness, independently of it, as well as the ability of our consciousness to imagine (visualize) this reality, if it exists. As we already know, our consciousness is capable of imagining anything, and even seeing what does not exist. The question remains as to how much what is seen is reality, and not just a product of our consciousness.

Why do Newton's and Coulomb's laws have such a simple form as the reversed square of distance? Because the area of the sphere over which the "energy of the source" is distributed is proportional to the square of the radius and, accordingly, weakens as the square of the distance as it moves away from the center. In addition, of course, the parameters included in the equation for "force", such as masses and charges, are correctly guessed.

Generally, with the right choice of variables, many of our

equations take a compact form. Conversely, an awkward form most often indicates a poor or inadequate choice of variables. For example, we can still consider that it is not the Earth that revolves around the Sun, but, on the contrary, the Sun revolves around the Earth. It doesn't change anything fundamentally, as it's just a frame of reference. But at the same time, it is clear that instead of the elegant Kepler laws, you will be forced to deal with much more complex equations, the meaning of which will be impossible to comprehend, due to the complexity of their solutions.

It is also known that instead of the very compact and elegant laws of thermodynamics, we could write down Newton's laws of motion for every particle of the thermodynamic system and, in principle, solve them for any unknown quantity. However, this solution would look horrific and would be a nightmare, even for the most powerful modern computers. The results would be next to impossible to interpret in any sensible terms, but it's doable.

Therefore, it is not that the laws of physics are simple and beautiful or complex and ugly because of some supernatural reason independent not only of us but of the material world itself. It has to do with our choice of description, choice of variables, and choice of fictions (concepts) by which we describe them. They are strictly a product of our consciousness.

From this I conclude that we are only at the very beginning of awareness of the world around us and its mathematical description. In order to move forward we must take into account the critical role that our consciousness plays in this awareness.

Finding an explanation for the observed constants, such as the speed of light, the fine structure constant, the mass and charge of the electron, and many others is the task of science, and not at all a reason to give up by saying that this is someone's "design" and that there are no other explanations. Eventually, all of these puzzles will have adequate explanations. Like the one about John winning the lottery.

Is motion fundamental?

This could be the craziest question of all. If we are to doubt the reality of motion, which we concluded above is probably the only fundamental quantity in this world, then what is left? Is there anything real at all? This is what I really wanted to discuss. It is not beyond the realms of possibility that motion itself is just a fiction. Otherwise, why is it always accompanied by a counter-motion? This basically means that its existence is not welcome in the motionless world in which our world is embedded.

How is it possible that motion is a fiction? This is very easy to imagine. Consider this picture of "running lights":

Fig. 5

Imagine the bulbs coming on for a short moment one by one, let's say starting from the left. The bulb to the right comes on only after the previous bulb turns off. Furthermore, if you are looking at it in the dark, so that you cannot see a bulb when it is off, then you have the impression that a light or lit object is moving by itself. Notice that there is no motion here at all, just bulbs turning on and off in a sequence. In this way, we can easily mistake imaginary motion for something that is really just a sequence of static events. Moreover, there could be no events at all. Imagine that the bulbs do not light up but just fall into our line of sight one by one in the same way. Consequently, we can easily mistake moving our line of sight for real motion.

Am I about to say that all the motion that we observe in the world could be as fictitious as the one I just described? I would like to, but I cannot. When you get hit by a rock thrown at you by someone you realize how real this motion is. However, when we only observe, it is impossible to distinguish between the real motion of a physical object and an imaginary sequence of events.

In addition, I am unsure what can be called "real motion" in the absence of the space and time we discussed earlier. How could there be motion through non-existent space? It is impossible. Then what is it that we see (and feel)?

The nature of things is the most significant question about the world around us. While physics has so far been quite successful in determining the relations between various quantities, it has not answered a single question about the nature of things. Perhaps these are not the questions for physics but rather for philosophy or the sciences of consciousness. Our ability to ever answer these questions remains a mystery to me.

Conclusions

1. The fundamental constituent of human consciousness is the concept. Biologically, a concept is a unique sensation encoded in an individual's nervous system as a representation of an image associated with an abstract entity.

2. "I" is just one of the concepts.

3. Information is a product of consciousness and is the meaning of a signal. The meaning, or interpretation, is realized as a conceptogram, or vector of concepts.

4. The production of information requires learning and conceptualization.

5. The emergence of a new concept is a phase transition.

6. The emergence of the apparatus of concepts became the birth of humans and information.

7. Language emerged, and continues to expand, as a result of the emergence of conceptual thinking and the need to label a huge number of new concepts.

8. Biologically, consciousness is a sensation of activity in the brain. Functionally, consciousness is vision.

9. Memory is the association of sensations with a certain image, thought, or concept.

10. All thoughts and thinking are visual. Thought is an image or set of images produced by the brain. Thinking is either searching for a particular thought-image or browsing through them. Language or verbalization is a facilitator of this process.

11. Nothing "immaterial" exists. Everything that seems immaterial to us is actually a product of consciousness. All concepts, including information, have a perfectly material, tabulated sensation in our brain (and body) accompanied by images, albeit different from person to person.

12. Homo sapiens (wise, thinking man) is who we may become in future if we realize that today we are still Homo falsus, i.e. the lying man.

13. We live in a world of events and motion, and only we can understand and study it. Space and time are not independent, fundamental entities, but are derivatives of events and motion.

14. Physics operates with fictions. All "conservation laws" of physics are the consequences of the law of nullification of fictions.

15. Quantum probabilities exist only in our minds. Probabilities in physical formulas mean only one thing: they describe not the real phenomenon but its representation in our consciousness.

16. There are no "laws of nature". There are laws of physics that we invented. Their simplicity or complexity is determined by how well we guess and how much the variables we use correspond to the reality that they are supposedly describing.

Afterword

To conclude, I'd like to call on all of us to stop lying. But this is useless, because it is the same as calling on people to stop talking. Therefore, I urge us all to start being more critical of what we say and what we think, as it all could be false.

Let me start with myself. Even though I hope that at least some of my thoughts presented here may be true or relevant to reality, I have absolutely no idea what it consists of, and whether it exists at all.

Acknowledgements

I am thankful to the late Professor N.N. Meiman (and the High Energy Physics group of Tel Aviv University for their hospitality), with whom I had the good fortune to share an office in 1991-1993 and talk about aspects of the scientific mind and the peculiarities of scientific thought. It was his collaborations and friendship with many famous physicists, such as Landau, Zeldovich, Sakharov, and Ginzburg that led to these observations. They provided me with invaluable insight into writing this text. I am grateful to A. Ushveridze for reading and discussing this manuscript. I am grateful to Lev Neiman for discussions and for his valuable comments. I am also indebted to Professor Vitaly Polunovsky for enlightening me on some aspects of modern genetics and for discussing various issues related to this text. I thank Alex Gorsky for reading the manuscript and for his valuable comments. I am indebted to Kevin McKinney for reading and correcting the English version of this manuscript. Finally, I am grateful to James Kingsland for his editorial advice and copy-editing.

Bibliography

[1]   Ann Gibbons, "Bonobos join chimps as closest human relatives", Science, 13 June 2012.

[2]   Rita Carter, "Mapping the Mind", book, 2010

[3]   T. Chernigovskaya, "How the brain works", 2011, and other talks

[4]   Daniel Dennett, "The Illusion of Consciousness", 2007, and other talks; "Consciousness Explained", book, 1992.

[5]   Chalmers, D. J. "Facing up to the problem of consciousness", Journal of Consciousness Studies 2: 200–219

[6]   Boris Rusakov, "Concepts as building blocks of information and the structure of consciousness", August 2022, unpublished.

[7]   Boris Rusakov, "Information is a product of consciousness", https://xpertnetinc.com/ai-project, October 2022.

[8]   Boris Rusakov, "Concepts as elementary constituents of human consciousness", October 2022, arXiv: 2208.09290, 2022.

[9]   C. Shannon, "A mathematical theory of communication", The Bell System Technical Journal, 27: 379–423, 623–656, July1948, October 1948.

[10]   MacKay, D. "Information, Mechanism and Meaning", MIT Press: Cambridge, MA, USA, 1969.

[11]   Robert K. Logan, "What is information? Why is it relativistic and what is its relationship to materiality, meaning and

organization, Information, 3: 68–91, 2012; doi:10.3390/info3010068

[12] Yuval Harari, "Sapiens: A Brief History of Humankind", book, 2015.

[13] S.Seung, "Connectome: How the Brain's Wiring Makes Us Who We Are", book, 2013.

[14] Richard Dawkins, "The Selfish Gene", book, 1976.

[15] Anatoly Protopopov, "The Treatise of Love, as it is recognized by awful bore", book, 2003.

[16] Yuval Harari, "Homo Deus. A Brief History of Tomorrow", book, 2016

[17] K. Anokhin, V. Avetisov, A. Gorsky, S. Nechaev, N. Pospelov, and O. Valba, "Spectral peculiarity and criticality of the human connectome", Physics of Life Reviews, 31,240–256, December 2019.

[18] A.B. Migdal, "Search for Truth", book, 1983 (in Russian)

[19] A. Tsvelik, "Life in the Impossible World", book, Ivan Limbakh: St. Petersburg, 2012 (in Russian).

[20] B. Rusakov, "Loop averages and partition functions in U(N) gauge theory on two-dimensional manifolds", Modern Physics Letters A, 5: 693–703, January 1990.

[21] E. Witten, "On quantum gauge theories in two-dimensions", Communications in Mathematical Physics, 141, 1991, 153–209

[22] D. J. Gross, "Two-dimensional QCD as a string theory", Nuclear Physics B, 400:161–180, 1993, hep-th/9212149

[23]   B. Rusakov, "Exactly soluble QCD and confinement of quarks", Nuclear Physics B, 507: 691–706, March 1997, hep-th/9703142.

About the author

Boris Rusakov earned a Ph.D. in theoretical physics and mathematics from the Academy of Sciences in the former Soviet Union, and for many years worked in high-energy physics (HEP) at various institutions around the world. His research areas were quantum field theory, quantum chromodynamics (QCD), lattice gauge theory, and two-dimensional gauge theories on Riemann surfaces. He is an author of highly cited publications in 2D QCD. His affiliations include:

Space Research Institute and Cybernetics Council at the Academy of Sciences, Moscow, Russia, 1988–1991.

HEP group at the Department of Physics and Astronomy, Tel Aviv University, Tel Aviv, Israel, 1991–1993.

HEP group at the International Centre for Theoretical Physics, Trieste, Italy, 1993–1995.

Physics Department, University of Oxford, Oxford, UK, 1995–1997.

Theoretical Physics Institute, Department of Physics and Astronomy, University of Minnesota, 1997–1999.

CERN – the European Organization for Nuclear Research, Geneva, Switzerland

Invited speaker at the universities of Princeton, Stanford, and Berkeley

In 1999, he founded IT consulting company and worked a consultant and a manager of various IT projects for Northern States Power, Xcel Energy, Siemens, and many others.

Since 2018, Rusakov has dedicated himself to researching various aspects of human intelligence, consciousness, and information. He has published papers about information as a product of consciousness, and concepts as its elementary constituents.

www.ingramcontent.com/pod-product-compliance
Lightning Source LLC
Chambersburg PA
CBHW071427210326
41597CB00020B/3682